SCIENCE SAMPLER
GRADES 4-8

WRITTEN BY SANDRA MARKLE
ILLUSTRATED BY BEV ARMSTRONG

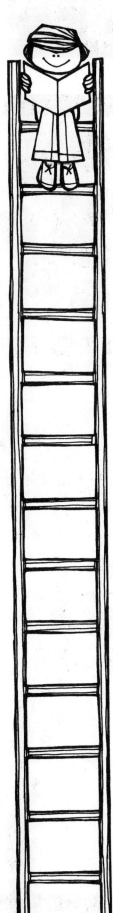

THE LEARNING WORKS
P.O. Box 6187 Santa Barbara, CA 93160

TABLE OF CONTENTS

To The Teacher	3
The Mysteries of Matter	4—21
Green Friends	22—41
Getting To Know Me	42—59
The Magic of Magnets	60—75
Mini-Beasts and Creepy Creatures	76—93
Air: The Invisible Force	94—112

The purchase of this book entitles the individual teacher to reproduce copies for use in the classroom.

The reproduction of any part for an entire school or school system or for commercial use is strictly prohibited.

No form of this work may be reproduced or transmitted or recorded without written permission from the publisher.

Copyright © 1980 — THE LEARNING WORKS
All rights reserved.
Printed in the United States of America.

TO THE TEACHER

The Science Sampler is designed to provide the teacher with a programmed approach to a challenging hands-on experience in science.

The book contains 6 units: matter, plants, senses, magnetism, insects and spiders, and air.

Each unit contains:

1. Brief background information for the teacher
2. A list of words that may be unfamiliar to the students
3. A specific set of learning objectives
4. Predictions to provide students opportunities to brainstorm, estimate results, and consider possibilities before they experiment
5. Display and bulletin board ideas to create the right atmosphere
6. Extra Spark Starters to light the fire of curiosity in your students
7. A list of materials so you will know ahead of time what is needed for all the experiments in each unit
8. Additional sources, both books and audio-visual materials
9. Answers to questions and puzzles
10. An evaluation section to see if students have mastered the learning objectives
11. Extended Learning suggestions for more creative and challenging ways to continue the unit and ideas for incorporating language arts, math, and social studies

The introductory story of each unit is also a learning motivator. The investigations are coded *I* for those that can be completed by students without any assistance, *TD* for experiments that may need teacher direction, and *A* for the more challenging experiments.

The Science Sampler's goal is to equip you, the teacher, with as many time-saving tools as possible so that you can enjoy sharing the challenges of these science units with your students.

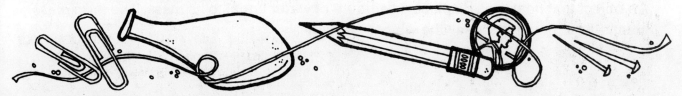

THE MYSTERIES OF MATTER

BACKGROUND

Everything is made up of tiny particles called *molecules* and *atoms*. These particles are constantly in motion. Depending on the speed of their movement and the spacing between them, different kinds of matter are formed.

In a *solid*, the particles are close together and moving very slowly. In a *liquid*, the particles are further apart and moving at a medium speed. In a *gas*, the particles are very far apart and moving rapidly.

There are other properties that are characteristic of the three kinds of matter. A solid has a definite shape and can be compressed only with a lot of pressure. A liquid has no definite shape and resists being compressed. Try pressing a puddle of water with your hand. The water squishes out rather than compacting into the smaller space. A gas has no definite shape but can be easily compressed. Scuba divers use tanks of compressed air.

One kind of matter may change to another type just by having the speed of its molecules or atoms change. If water molecules are speeded up by the addition of heat energy, they spread apart, forming a gas. If water molecules are slowed down by the subtraction of heat — cooling — they move closer together, forming ice.

When solids form, the atoms arrange themselves into geometric shapes called *crystals*. A crystal for a certain solid always takes the same shape. A crystal of salt, for example, is always a cube.

Students should use all of their senses to examine examples of matter. They should understand, however, that no unknown substance should ever be tasted, no matter how harmless it appears.

There is also a safe way to smell unknowns. Fan your hand over the substance to be smelled. If there is any odor, a little of it will be carried to your nose. The smell won't be as strong as if you leaned close and sniffed, but neither will there be any danger of inhaling a substance that could damage the nasal tissues.

All matter occupies space. The question is how much space does matter take up? An object that floats will displace an amount of water equal to its mass. The term *mass* means how much material the object contains.

Copyright © 1980 — THE LEARNING WORKS

The question of mass is what concerned Archimedes when he tried to find out if the king's crown was pure gold or a mixture of gold and another metal. The story goes that Archimedes figured out the solution to this question while taking a bath. He noticed that the water rose when he got into his tub. He then had a pure gold crown made that was equal in mass to the king's crown. When he tested them, the two crowns displaced different amounts of water. Therefore the two crowns did not have the same density.

EUREKA! The king's crown was not pure gold.

Gravity is the force that is constantly pulling everything toward the center of the earth. Every object of matter has a center point in its mass, its center of gravity. This center point has to be taken into consideration to build buildings, construct mobiles, and even balance the human body.

WORD BOX:

properties	matter	molecule
liquid	solid	gas
crystal	saturated	solution

LEARNING OBJECTIVES:
1. Students will be able to identify solids, liquids, and gases.
2. Students will be able to tell the main properties of each form of matter.
3. Students will recognize that matter takes up space in proportion to its size and weight.
4. Students will understand how crystals form.
5. Students will recognize how the shape and arrangement of a solid affects its strength and balance.

PREDICTIONS:
1. How do you think that Archimedes could find out if the crown was really pure gold? (after reading the introductory story)
2. Which do you think will displace more water, the small block of wood or the golf ball? Why?
3. Which do you think can support more weight, the plain sheet of notebook paper or the folded sheet of paper?

DISPLAY AND BULLETIN BOARD IDEAS:
1. Divide a display or bulletin board area into three parts. Label the sections solid, liquid, and gas. Have a picture in each section that shows the arrangement of the molecules characteristic of that kind of matter.

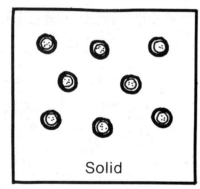
Solid

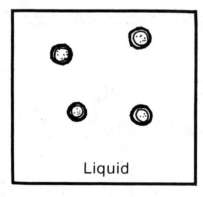
Liquid

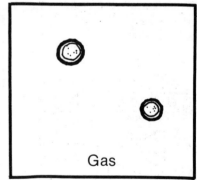
Gas

Put up pictures of examples of each kind of matter. Invite students to add pictures of examples that they find. Also include word cards of words that describe each type of matter. For example, *hard* would describe a solid, *wet* would describe a liquid, and *misty* would describe a gas. How many more words can you and your students think of to describe the three kinds of matter?

2. Have a Touch Table with different kinds of matter to feel.

3. In a special display spot, put out a mystery sack. Put something in the sack, perhaps wrapped in paper so that it cannot be felt through the bag, and staple the bag shut. Each day put out a one-word clue about the object in the mystery sack. Can your students identify what kind of matter the mystery object is from the clues? Can they identify the object?

EXTRA SPARK STARTER:

1. Make molecule models of a solid, a liquid, and a gas using clay and plastic straws (or salt dough and toothpicks, bake and paint).

2. Divide the class into three teams after an initial discussion of the properties of a solid, a liquid, and a gas. Have slips of paper that say *solid* or *liquid* or *gas.*

   Play the game like charades with the difference that each team in turn must act out the kind of matter that they represent so that the rest of the class can guess it correctly. Each team scores five points each time that they are successful.

3. Another team activity can help students understand in which category different objects belong. Depending on the size of the class, divide the students into teams.

   Give each team leader three loops of colored construction paper — one ring each for solid, liquid, and gas. Then equip the teams with construction paper, scissors, glue, and marking pens, and challenge them to see who can attach the most rings to each category. One object can be listed on each ring, but it must fit unquestionably into that category of matter. For instance, wood is a solid, water is a liquid.

   Put a time limit on this activity and then check the results with the entire class. Each team scores one point for each correctly placed object. The finished paper chains make nice room decorations, and each time your students look at them they will be reminded of just what objects are solids, liquids, and gases.

MATERIALS NEEDED:

Rock
Tall glass
Jar with a lid
Metal pan
Kitchen scales or triple beam balance scale
Paper clips
Shallow dish
Laundry bluing
Styrofoam cups
2 potatoes
Bottle with cork
Screwdriver
4 metal nuts (each weighing about 2 ounces)

Cotton balls
Bottle of soda pop
Masking tape
Metric ruler
Pie pan
Tablespoon
Wooden spoon
Cotton string
Ammonia
Food coloring
Pennies
Knife
Pencils
1 cm (½ inch) elastic or a piece of old inner tube

Rubber eraser
Balloon
Golf ball
Small block of wood
Small saucepan
2 cups of granulated sugar
4 or 5 charcoal briquettes
Salt
Notebook paper
Pebbles
5 forks
3 lb. can (empty) with plastic lid
2 straight pins

ADDITIONAL SOURCES: BOOKS

Berry, James. Exploring Crystals. Crowell, 1969.

Challand, Helen J., and Elizabeth R. Brandt. Science Activities From A to Z. Grosset and Dunlap, 1963.

DeVries, Leonard. The Second Book of Experiments. The Macmillan Co., 1968.

Goldstein-Jackson, Kevin. Experiments With Everyday Objects. Prentice-Hall, Inc., 1978.

Herbert, Don. Mr. Wizard's Science Secrets. Hawthorn Books, Inc., 1965.

Stone, Harris, and Bertram M. Siegel. Puttering With Paper. Prentice-Hall.

ADDITIONAL SOURCES: FILMSTRIPS

Matter and Energy. Primary and intermediate level. Set of 6 filmstrips and sound. SVE-Society For Visual Education, Inc.

ANSWERS:

p. 11:

SOLIDS	LIQUIDS	GASES
Rock	Water	Air in balloon
Rubber eraser	Soda pop	Air in jar
Jar and lid		
Glass		
Cotton balls		
Bottle		

What happens:
1. Solid has a definite shape, molecules close together and cannot be compressed easily.
2. Liquid has no definite shape, molecules spread somewhat and cannot be compressed easily.
3. Gas has no definite shape, molecules spread far apart and can be easily compressed.

Puzzling: Answers will vary.

p. 12: On the chart and what happens, the answers will vary.
Puzzling: The water level increased because the objects took up space.

p. 13: Chart will vary.
Puzzling: An object displaces an amount of water that is equal to its mass. The mass of an object includes both its weight and how much space it takes up. Therefore, a very large, light-weight object could displace as much water as a small, heavy object.

p. 16: What happens:
1. The folded paper has a better support.
2. The folded paper can hold weight.
Puzzling:
Think about corrugated cardboard. This design gives the paper more strength. The ridges, like the folds, give more area of support.

p. 18: Archimedes can't bend over with his heels against the wall and his feet together. As he started to bend over, his weight would be beyond his center of gravity. The middle picture shows what would happen.

The mobiles balance well because the weights can be arranged around each object's center of gravity.

p. 19: 2. The can will roll back to you.
3. The weights wind up the elastic or rubber strip inside the can. When the strip is wound up tight, the can rolls backward as the strip unwinds.

EVALUATION:

Choose the letter of the word or words that correctly completes each statement. Print the letter on the blank. The answers may be used more than once.

A. Slowly   C. Gas   E. Solid   G. Rises
B. Liquid   D. Mass   F. Gravity   H. Crystals

1. __(E)__ is a kind of matter with a definite shape.
2. __(B)__ is a kind of matter with no definite shape but it resists being compressed.
3. When an object is put into water, the water level __(G)__ .
4. An object displaces an amount of water equal to its __(D)__ .
5. When solids form, they arrange themselves into geometric shapes called __(H)__ .
6. __(C)__ is a kind of matter with no definite shape and it can be easily compressed.
7. __(F)__ is the force that pulls everything toward the center of the earth.
8. How big and how perfect a crystal is depends on how __(A)__ it forms.
9. __(E)__ is a type of matter with its molecules tightly packed together.
10. __(C)__ is a type of matter with its molecules far apart.

EXTENDED LEARNING:

1. Find out more about Archimedes and his experiments.
2. Put on a Matter Fair for other classes. Present experiments about matter.
3. Think of ways to change matter: tear, crumple, write on, burn, etc.
4. Do more origami.
5. Write about what it would be like to be able to change into another form of matter. What if you could become invisible?

Name _____

THE MYSTERY OF THE KING'S NEW CROWN

Long ago in Greece there lived a very rich king. One day the king decided that he wanted a new crown. He sent for his goldsmith.

"Use this gold," the king told the goldsmith, "and make me a crown of pure gold."

The goldsmith measured the king's head, took the gold and left.

When the crown was finished, the goldsmith brought it to the king. The crown was beautiful, but the king was not happy.

The more the king looked at the crown, the more he became sure that the goldsmith had cheated him. The crown felt heavy, but the king had a feeling that the crown was not <u>pure</u> gold.

If the king had understood the mysteries of matter, he could have found out for himself.

The king called Archimedes, a very wise mathematics teacher, to help him solve the mystery.

In these experiments, you will investigate matter. Keep the king's problem in mind. See if you can figure out how Archimedes could tell if the king's crown was pure gold or only a mixture of metals.

Name _____

WHAT IS MATTER?

Supplies needed:	Rock	Cotton balls
	Rubber eraser	Bottle of soda pop
	Glass of water	Balloon full of air
	Jar (full of air) with a lid

How to do it:

1. Use your eyes, your ears, your nose, and your fingers to examine each sample, including the air in the jar. Listen to the air in the balloon.

2. After testing, decide whether each sample is a solid, a liquid, or a gas.

3. Write the name of each sample on the chart.

What happens:

1. Tell two ways that a solid is different from a liquid or a gas.

2. Tell two ways that a liquid is different from a solid or a gas.

3. Tell two ways that a gas is different from a solid or a liquid.

Puzzling out the results:

Be a good scientist. Find all the things you can that are solids, liquids, and gases to fill in the chart.

Solid	Liquid	Gas

Make a poster showing pictures of solids, liquids, and gases.

Copyright © 1980 — THE LEARNING WORKS

Name _____

DOES MATTER TAKE UP SPACE?

Supplies needed: Tall glass   Metric ruler
                  Masking tape   Rock
                  Metal pan   Small block of wood
                  Golf ball

How to do it:

1. Fill the glass half full of water and put a piece of tape on the glass to mark the water level.

2. Put the glass on the pan in case the water runs over.

3. Carefully put the golf ball in the water (do not drop it).

4. Using the ruler, measure how much higher the water is above the tape level. Record this measurement on the chart.

5. Repeat using the rock and the block of wood. Record the increase in water level for each object.

Objects	Increase in water height

What happens:

Try this experiment at home in the bathtub. Fill the tub halfway with water. Mark the level with tape, and then get in.

How much do you raise the water level in the tub? _____

Puzzling out the results:

Why do you think there seems to be more water in the glass when the objects are added?

I, TD   Name _____

HOW MUCH SPACE DOES MATTER TAKE UP?

Supplies needed: Golf ball   Kitchen scales (triple beam
                               Pie pan   balance scale if
                               Tall glass   available)

How to do it:

1. Place the golf ball in the pie pan. Weigh the ball-pan combination on the scales.

2. Write the weight under the picture of the golf ball and pan.

3. Place the glass on the pan and fill it to the very top with water.

4. Put the ball in the water (do not drop in).

5. When the water has stopped overflowing, carefully remove the glass and the ball. Do not spill more water as you move the glass and ball.

6. Weigh the pan and water. Write the weight under the picture of the pan and water.

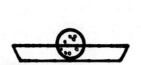      

_____   _____

If you worked carefully, the weight should be nearly the same both times. If your results were very different, try again.

Puzzling out the results:

Why do you think the weight of the water that spilled over was nearly the same as the weight of the ball?

A

Name _____

EXPERIMENT FOR YOUNG WIZARDS

Everything is made up of tiny particles called *atoms* and *molecules*. When solids form, these particles often arrange themselves into geometric shapes called *crystals*.

Whenever a crystal of a certain solid forms, it always takes the same shape. A crystal of salt is always a cube. A crystal of water — a snowflake — always has six sides.

How large and how perfect crystals are depends on how slowly they form. When a liquid changes slowly to a solid, the atoms have time to line up in new, orderly crystal shapes.

Here are two experiments for you to grow crystals of your own.

1. Rock Candy (A Tasty Experiment)

Supplies needed:
- ½ cup water
- Small saucepan
- 2 cups of granulated sugar
- 1 tablespoon
- Wooden spoon
- 1 tall glass
- 1 clean paper clip
- 1 piece of clean cotton string slightly less than the height of the glass
- Pencil

How to do it:

1. Put the half cup of water in the pan and heat until hot (not boiling).
2. Add the sugar one tablespoon at a time.
3. Stir with the wooden spoon after each addition.
4. Do this until the sugar no longer dissolves when you add it (it should take most of the two cups).
5. Heat and stir until the sugar solution boils.
6. Boil one minute or until the solution is thick and clear.
7. Let the solution cool, and then pour it into the glass.
8. Tie the paper clip to the string and attach the string to the pencil.
9. Place the clip in the sugar solution and let the solution sit at room temperature for a week or more.
10. If the top of the solution crusts over with sugar, break the crust gently.

What happens:

1. How long does it take for the first rock candy crystals to form? _____

A   Name _____

2. Draw a picture of the sugar crystals after one week.

If no crystals form after one week, you didn't dissolve enough sugar in the solution. Reheat and try again.

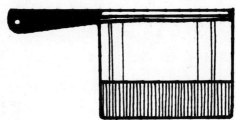

2. A Crystal Garden

Supplies needed: 4 or 5 charcoal briquettes   6 tablespoons of salt
                   Shallow dish or fish bowl   6 tablespoons of water
                   1 tablespoon of ammonia   6 tablespoons of laundry bluing
                                                   Food coloring

How to do it:

1. Place the charcoal briquettes in the dish or bowl.

2. Combine all the other ingredients except the food coloring in a bowl and mix well.

3. Pour these ingredients over the charcoal.

4. Sprinkle on several drops of food coloring.

5. Let your garden sit at room temperature for a week or more.

What happens:

1. How long does it take for the crystals to begin to grow? _____

2. Draw and color a picture of your crystal garden.

Name _____

IS THE SHAPE OF A SOLID IMPORTANT?

Supplies needed: 2 pieces of notebook paper
 6 styrofoam cups
 12 pennies or pebbles

How to do it:

1. Place one sheet of notebook paper between two cups flat like a bridge.

2. Put a third cup on top of the paper.

3. Fold the second sheet of paper accordion style.

4. Place the folded paper between two styrofoam cups like a bridge.

5. Put a third cup on top of the folded paper.

What happens:

1. Which paper was a better support? _____

2. One by one, add pennies or pebbles to the cups on top of the pieces of paper. Which paper can support more weight? _____

Puzzling out the results:

Why do you think the shape of the solid made a difference?

Building materials are designed to give as much support as possible with the the lightest weight.

Copyright © 1980 — THE LEARNING WORKS

I, TD   Name _____

PUZZLE FUN

Just for fun — change a flat piece of paper into a cup that you can drink from.

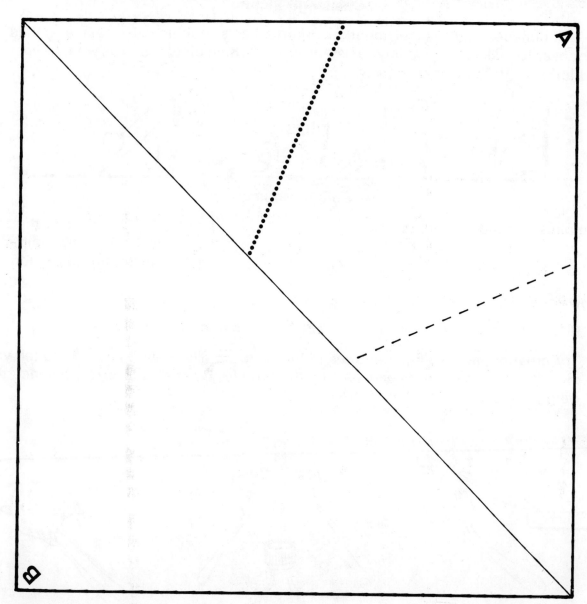

1. Cut out the square and fold along the solid line so that A is on top of B.

2. Fold the right corner up along the — — — line and over so that it meets the edge of the paper.

3. Fold the left corner up along the ••••••••• line, and over on top of the right corner.

4. Fold the top points down so that one is on either side of the cup. Squeeze the cup carefully to open it. Have a drink!

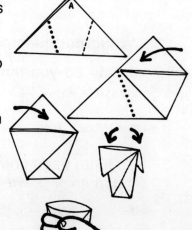

Name _____

HOW DOES GRAVITY AFFECT MATTER?

Challenge Picture

Archimedes had his feet together and his heels touching the palace wall as he bent over to pick up something he had dropped. Which picture shows what happened to him? Try it yourself to find out.

Potato Mobiles

Supplies needed: 2 potatoes   1 bottle with a cork in the top
                                Knife   3 pencils (with sharp points)
                                5 forks   1 piece of string 30 cm long

How to do it:

1. Slice one raw potato into rounds about one cm thick.
2. Try creating mobiles like the ones in the pictures. There isn't any way to tell you exactly where to place each part of the mobile. By trial and error, balance the weight around the center of gravity.

 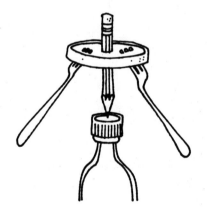 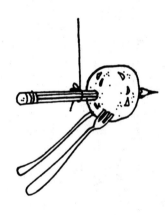

Puzzling out the results:

    Why do you think these mobiles balance so well?

Matter has a point around which all its weight is centered. This point is called the *center of gravity*. Think where the center of gravity is in each mobile.

    Can you create other mobiles of your own?

A

Name _____

EXPERIMENT FOR YOUNG WIZARDS

In this experiment, you will have another chance
to explore how gravity affects matter.

Supplies needed:

3-pound can with a plastic lid (such as shortening and powdered drinks come in)
Screwdriver
1 paper clip

1 piece of 1 cm (½ inch) elastic or a piece of old inner tube slightly longer than the length of the can
4 metal nuts (each weighing about 2 ounces)
2 straight pins

How to do it:

1. Mark a spot in the center of the bottom and the center of the lid of the can.

2. Use the screwdriver to punch a hole about two centimeters wide through each mark. (Turn the can over when you do the bottom so that you punch toward the inside. This keeps the sharp edge inside. Be careful.)

3. Unbend the paper clip to form an "S"-shaped hook.

4. Hook one end of the paper clip through the center of the elastic strip. Hook as many of the nuts as you can over the other end of the clip. Bend the end up to hold the nuts in place.

5. Slide the elastic strip through the slot in the bottom of the can. Put a pin through the elastic strip to keep it from sliding back out.

6. Put on the lid. Slide the other end of the elastic strip through the slot in the lid.

7. Pull the strip tight and hold it in place with another pin.

8. If you could see through the can, it should look like this:

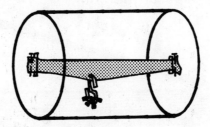

What happens:

1. Roll the can away from yourself on a smooth floor.

2. What does the can do? _____

3. Why do you think the can reacts this way? _____

_____

Copyright © 1980 — THE LEARNING WORKS

Name _____

PUZZLE FUN

When Bonanno Pisano built the Tower of Pisa, he didn't make the foundation deep enough to reach solid ground. The building also was not wide enough to support the tower's weight.

After the first few floors were built, Bonanno Pisano noticed that the tower was leaning. He tried to build the next few floors leaning the other way to correct the problem.

Pisano's attempt failed, and the city officials fired him. No one has been able to correct Pisano's mistake. The tower has been sliding over its center of gravity at the rate of about sixteen millimeters per year.

Get a stack of pennies. Each penny will represent a floor of the Tower of Pisa. Build a tall tower on the edge of a book.

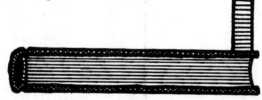

How far can you lift the edge of the book before the tower leans so far that it falls over?

If you have a protractor, have a friend help you measure the angle when the tower collapses. Try again. Does the tower always collapse at the same angle?

If you think of a way to stop the Leaning Tower from falling, write your idea to:

The Italian Government Travel Agency
630 Fifth Avenue
New York, NY 10020

Archimedes gave the king's problem a lot of thought. Did you? Archimedes finally figured out the answer while he was taking a bath.

How do you think Archimedes helped the king find out if his crown was *pure* gold?

THE MYSTERIES OF MATTER

_____

Has successfully mastered THE MYSTERIES OF MATTER

Date: _____

GREEN FRIENDS

BACKGROUND

There are many different kinds of plants in the world. The experiments in this section are devoted to the study of green plants.

All green plants contain a special substance called *chlorophyll*. Chlorophyll is an important part of a process called *photosynthesis*. The word *photosynthesis* means "to make with light." Only green plants are capable of making their own food.

Scientists are not sure exactly how plants are able to do this, but they do know that photosynthesis follows this formula:
Sunlight + Water + Chlorophyll + Carbon dioxide = Sugar and starch + Oxygen.

Any part of the plant that is green can make food, but the leaves are the main food factories. Therefore, plants by necessity have to have a system to carry water up from the roots to the leaves and food down to be stored in the roots and the stem.

The *xylem* tubes carry fluids upward in plants and the *phloem* tubes carry fluids down. The downward movement is controlled mainly by gravity but *transpiration* and *capillary action* are needed to move water up to the leaves.

Transpiration is the process of water evaporating from the leaves. This process, combined with natural growth and photosynthesis, creates a pull that begins moving water upward in the plant.

The water then moves cell by cell — just as it rises in a dry paper towel when one corner is dipped in water. Water molecules tend to cling together and so a whole column of water rises. This is capillary action. It is a powerful force. Think how far the water has to rise in a giant redwood tree.

Through photosynthesis, plants also exchange carbon dioxide for oxygen. Without plants, we would run out of oxygen to breathe. Plants are not totally oxygen producers, however. When it is dark, plants use the food that they have produced to grow. During growth, oxygen is used up and carbon dioxide is released into the air. Gas exchange is again mainly through the leaves.

Plants are also much more able to move than many people realize. Leaves and flowers turn toward sunlight. The head of a sunflower follows the sun across the sky. Roots move toward water. The tip of the root produces an acid that allows the root to push through most obstacles. In time, roots can even break through cement. These movements are called *tropisms*.

Green plants that grow from seeds only need moisture to begin growing. As the seed sprouts, the root pushes out first. Then the stem with its first leaves appears. The part that would be food for us if we ate the seed is called the *cotyledons*. These cling to the stem and supply the young plant's energy for growth during the first couple of weeks.

After that time, to grow well, the plant needs sunlight, an adequate supply of water, and loose soil rich with minerals. If any of these needs is not met, the plant's growth will be stunted or the plant may die.

WORD BOX:

sprout	cotyledon	xylem
capillary action	photosynthesis	transpiration
tropism		

The underlined words are not used in the experiments but the processes are explained.

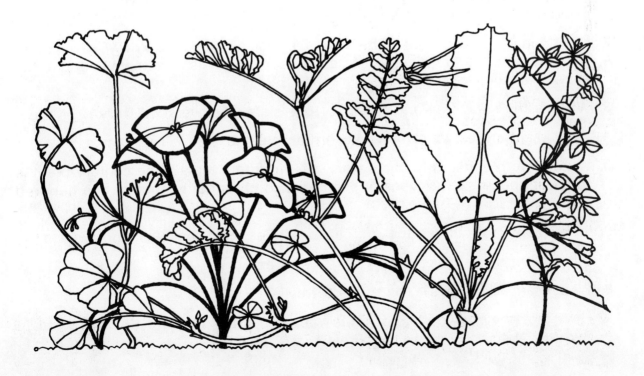

LEARNING OBJECTIVES:
1. Students will understand how a seed sprouts.
2. Students will be able to choose the best kind of soil for plant growth.
3. Students will understand the importance of sunlight to the growth of green plants.
4. Students will be able to supply an appropriate amount of water for plants.
5. Students will understand capillary action.
6. Students will be able to state the important relationship between plant life and animal life.

PREDICTIONS:
1. Can plants sprout without being in dirt?
2. Can plants grow in total darkness?
3. (Show sand, potting soil, and clay) Which kind of soil do you think will grow the best plants? Why?
4. Can a plant move?

DISPLAY AND BULLETIN BOARD IDEAS:
1. Put up a brightly colored bulletin board with pictures of beautiful and unusual plants.
2. Fold sheets of plain white paper in half lengthwise. Print questions about plants on the outside and the answers on the inside. For older students you may want to print the questions and provide reference books for them to find the answers on their own.

Here are some suggestions for a MEAN GREEN CHALLENGE.
1. What plants live the longest? There are a number of long-living plants. One of the oldest is the bristlecone pine, estimated at over 4,000 years old.
2. What plant catches flies? The Venus flytrap is the best known flycatcher. Sundews and pitcher plants also catch insects and dissolve them to add to the mineral content of their diet.
3. What is the state tree and flower of our state? Answers will vary.
4. What plant stores water in an accordion-pleated stem? The cactus.
5. What plant blooms only once every one hundred years? The century plant is the best known. Some types of bamboo also bloom only once in one hundred years.
6. What plants roll across the ground to scatter their seeds? Tumbleweeds.
7. What plant helps make a car? A rubber tree.
8. What plant has knees? The cypress tree.
9. How big is a redwood tree? This varies, but the General Sherman tree in Sequoia National Park in California is 272 feet tall.

Copyright © 1980 — THE LEARNING WORKS

EXTRA SPARK STARTER:

1. Make a chart of the predictions. Discuss them. Take a vote on the answers. List these on the chart. Tell students that after they have had a chance to experiment, you will return to the chart to see how many people were right.

2. This is a good time to brainstorm. Is there anything that your students would like to find out about plants? This could be included on the chart for later discussion. If the question is not covered by the experiments, it could be investigated later by the class or by volunteers.

3. Soak bean seeds overnight. Give each child one or two seeds to investigate.

    1. What part do we eat?

    2. How does water get into the seed?

    3. Can you find the embryo — the young plant?

    4. How is the embryo like a full-grown plant? How is it different?

    5. What do you think is the job of the two large halves of the bean seed?

MATERIALS NEEDED:

Large sponge
*Bean seeds
Plastic wrap
Metric ruler
Potting soil
Clay soil
Clean sand
Paper plates or aluminum pans
Knife or razor blade
2 tall glasses
Long-stemmed, white flower (carnation)
Wood toothpick
2 pill bottles
Celery with leaves
Medium-size box with lid or flaps

½ teaspoon
Tablespoon
Teaspoon
*Sunflower seeds
*Radish seeds
*Tomato seeds
*Grass seeds
3-quart dish or pan
5- or 10-gallon fish tank
Dry fertilizer
Styrofoam cups
Bean plants (grow from seeds in styrofoam cups until about 10 cm tall)
Sprouting potato

Food coloring
Small pot
Stiff cardboard
Matches
3 clip clothespins
**Water weeds (such as Anacharis)
Plastic funnel
Lemon juice
White paper
½ plastic egg
Permanent magic markers or acrylic paint and brush
Masking tape
Modeling clay
Scissors
Construction paper
Crayons

*Other seeds or dried seeds from grocery store may be substituted if seeds listed are not available. Bird seed could be a source for seeds.

**Can be collected or purchased in any store that sells aquarium supplies. Elodea is another good water weed to use.

ADDITIONAL SOURCES: FILMSTRIPS

Ecology: Interrelationships In Nature. Intermediate level. Set of 8 filmstrips and sound. SVE-Society For Visual Education, Inc.

Exploring Ecology. Grades 5-12. Set of 5 filmstrips and sound. National Geographic Society.

Learning About Ecology. Primary and intermediate level. Set of 6 filmstrips and sound. Encyclopedia Britannica Corporation.

The Seed Plants. Grades 4-6. Set of 8 filmstrips and sound. Coronet.

Trees. Grades 4-6. Set of 6 filmstrips and sound. Coronet.

ANSWERS:

p. 29: 1. The seed coat splits open.
      2. The root pushes out first.
      3. Answers will vary, but it will be about 1-3 days. Some seeds sprout within hours.

p. 30: Puzzling:
Answers will vary but the answers will probably not be over two weeks.

p. 31: Plants should grow best in the potting soil, providing they receive the proper amount of water and sunlight. Sandy soil allows the water to drain away too quickly. Clay holds the water next to the roots and can cause them to rot.

p. 32: Challenge Picture: The plant will be long and skinny (middle picture) because it is trying to grow to reach the sunlight.

p. 33: Puzzling:
The plants were trying to grow toward the sunlight. The plants in the dark were white or yellowish-white. Without sunlight the chlorophyll — coloring that makes the plant green — doesn't develop.

p. 34: Too much water, like too little water, can kill a plant. Measurements will vary.

p. 35: What happens:
      1. The length of time will vary. This experiment can be done with any vining plant. Beans can be used by letting them sprout in the dark box.
      2. The potato sprout is pale because without the stimulation of sunlight, the plant does not produce chlorophyll.

Puzzling:
Plants are able to move. Plants react particularly to water and sunlight. Roots move toward water. Stems and leaves move toward light. These movements are called tropisms.

p. 37: Puzzling:
Water vapor is constantly escaping from the leaves. As the water evaporates — becomes part of the air — it creates a pressure inside the plant similar to sucking on a straw. Water molecules all tend to stick together, so the water rises in the plant.

p. 38: What happens:
1. The stick bursts into flame.
2. The plant has given off oxygen.

p. 39: Puzzling:
The plants use the carbon dioxide given off by the animal life of the lake or pond and give off oxygen, which the animals need to live.

Trees and plants growing on our earth supply us with a constant supply of oxygen. Surprisingly, the biggest source of oxygen is the algae and seaweeds that grow in the ocean.

EVALUATION:
Fill in the blank with the word or words that would correctly complete each statement.

1. Only __(green)__ plants can make their own food.
2. When a seed sprouts the __(root)__ pushes out first.
3. Seeds have to have __(water)__ to sprout.
4. Most food for the plant is made in the __(leaves)__.
5. A plant grown in the dark will get __(tall and skinny)__.
6. Sand is not good soil for plant growth because __(the water runs through)__.
7. Clay is not good soil for plant growth because __(the water does not drain)__.
8. __(Xylem)__ tubes carry water up to the leaves.
9. __(Phloem)__ tubes carry food down to be stored.
10. Plants give off __(oxygen)__, which animals need to live.

EXTENDED LEARNING:
1. Put up a big U.S. map. Find out the state flower and tree for each state.

2. Adopt a tree and study it in all of the seasons.

3. Find out what products are made from plants.

4. Write a play about the adventures of the plant people.

5. Make a window greenhouse and grow plants to sell or to be given as gifts to parents and grandparents.

6. Have a visit from a botanist or florist. Visit a greenhouse.

7. Make terrariums.

THE SUPER SEEDS

When Jack's mother sent him to sell the family cow, he traded the cow for a handful of magic beans.

Jack's mother was very angry. "Beans," she yelled. "You traded our cow for beans!" She threw the seeds out the window.

During the night, the seeds sprouted and began to grow. Up, up, up — the young bean plants grew into the clouds.

When Jack got up the next morning, he was surprised to see the tall bean plants. Jack decided to climb the towering plants to see what was above the clouds.

What if Jack had known more about plants? With the proper soil, the right amount of water and some careful tending, Jack might have grown the world's biggest and best beans.

Perhaps these bean plants would have been a bigger treasure than the one Jack got from the giant.

In these experiments, you will investigate plants. Think about how Jack could have made the most of his special bean seeds.

Name _____

HOW DO SEEDS SPROUT?

Supplies needed:
- 1 large sponge (real or artificial)
- 1 3-quart shallow glass dish or small fish tank
- 10 bean seeds (dried beans)*
- 10 sunflower seeds*
- 10 radish or grass seeds*
- 10 tomato seeds*
- Plastic wrap
- Metric ruler

*Seeds sprout best if soaked in tap water overnight.

How to do it:

1. Rinse the sponge well in clean water and squeeze out.

2. Place the sponge in the glass dish or fish tank.

3. Lay one or two rows of each of the kinds of seeds across the sponge. The seeds may touch but should not overlap.

4. Cover the glass container with plastic wrap. On the outside of the glass, put a piece of masking tape with the date the seeds were planted.

5. Place the glass container in a warm, sunny spot.

6. Check the sponge every couple of days. Sprinkle on water to keep the sponge damp.

What happens:

1. What happens to the thin, tough coating on the outside of a seed as it starts to sprout?

2. What pushes out of a seed first — the root or the stem? _____

3. How many days does it take for each of the seed types to sprout?

Bean _____ Sunflower _____ Radish or grass _____

Tomato _____ Other (list type and tell number of days)

4. How many seeds never sprouted?

Bean _____ Sunflower _____ Radish or grass _____

Tomato _____ Others (list type and tell number)

Copyright © 1980 — THE LEARNING WORKS

Name _____

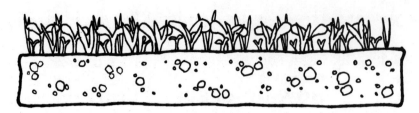

5. Draw a picture of one of each kind of plant two days after the leaves first appear. Measure the plant you draw and record the height below the picture.

PLANT'S FIRST BABY PICTURE

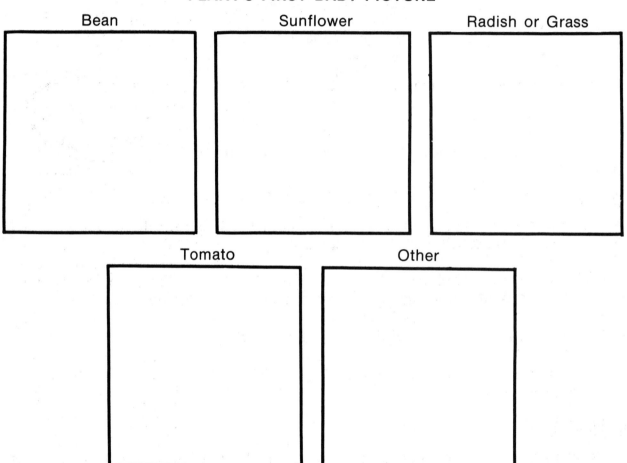

Puzzling out the results:

   Your seeds were able to sprout without being in dirt. They started on food stored in the seed.

   If you want your plants to continue to grow, transplant them into pots with rich dirt or potting soil.

   Leave one or two of each type of plant on the sponge. How many days do the plants continue to grow on their stored food? _____

Copyright © 1980 — THE LEARNING WORKS

Name _____

HOW IMPORTANT IS SOIL TO PLANT GROWTH?

Supplies needed:
- 24 bean seeds (dried beans)
- 8 styrofoam cups
- 2 cups of potting soil
- 2 cups of clay soil
- 2 cups of clean sand
- Dry fertilizer (follow directions on the package)
- 8 paper plates or small aluminum pans
- Metric ruler

How to do it:

1. Soak the bean seeds in tap water overnight.
2. Carefully poke two holes in the bottom of each cup for drainage.
3. Fill the cups according to the pictures. Label each cup. Plant three bean seeds in each cup.

A	B	C	D	E	F	G	H
SAND	½ SAND + ½ CLAY	½ SAND + ½ SOIL	CLAY	½ CLAY + ½ SOIL	CLAY + FERTILIZER	SAND + FERTILIZER	SOIL + FERTILIZER

4. Water each cup with two teaspoons of water daily.
4. Put the planted cups on the plates or pans in a warm, sunny spot.

What happens:

1. Write the date on the chart when the first sprout appears in each pot.
2. Measure the tallest plant in each cup two days after sprouting. Measure that plant for five days. Record its height each day on the chart.

Sprout	A	B	C	D	E	F	G	H
Date								
1								
2								
3								
4								
5								

Puzzling out the results:

In which kind of soil do the seeds sprout first and grow best?

_____

Why do you think the seeds grew best in this kind of soil?

_____

DO PLANTS NEED SUNLIGHT TO GROW?

Supplies needed: 6 styrofoam cups   3 aluminum pans or
                  6 cups of potting soil   paper plates
                  24 bean seeds (dried beans)   Metric ruler

Remember, seeds sprout best if soaked in tap water overnight.

How to do it:

1. Carefully poke two drainage holes in the bottom of each cup.
2. Fill the cups with potting soil and plant four seeds in each.
3. Sprinkle water on the soil to moisten it.
4. Place two cups on a pan in a warm, sunny spot.
5. Place two cups on a pan in a warm, shaded area.
6. Place two cups on a pan in a warm, dark spot (can be in a closet or cupboard).
7. Check each day. Sprinkle with water as needed to keep soil moist.

CHALLENGE PICTURE

Circle the picture that you think shows
how the plants in the dark will look.

What happens:

Fill in the chart to show what happens
as your plants sprout and start to grow.

Light	Date Sprouted	Height of tallest plant			
		2 Days	4 Days	6 Days	8 Days
☀ (sun)					
⛅ (partial sun)					
⬛ (dark)					

Copyright © 1980 — THE LEARNING WORKS

Name _____

Puzzling out the results:

Why do you think that the plants in the dark were so much taller than the other plants?

_____

_____

What color were the plants in the dark? _____

Why do you think they were this color? _____

_____

Plants need sunlight to make *chlorophyll*. Chlorophyll is the green coloring in plants and helps them to make their own food.

JUST FOR FUN

1. Collect some pretty leaves or interesting plants.

2. Press flat between newspapers weighted down by books.

3. Arrange plants or leaves in a design on construction paper.

4. Use a sponge to dab tempera paint over the plants and the paper. Then carefully remove the plants.

5. Or — use a brush or sponge to apply paint to leaves. Press the leaves on construction paper. This will make a print. Do not use leaves that are dry and crackly for this project.

I, TD

HOW DOES WATER AFFECT PLANT GROWTH?

Supplies needed: 4 healthy bean plants (you can use plants from an earlier experiment or plant bean seeds and use plants that are about 4 cm tall)   4 small aluminum pans or paper plates
½ teaspoon
1 teaspoon
1 tablespoon
Metric ruler
Crayons

How to do it:

1. Place one plant on each pan and put all the plants in a warm, sunny spot.
2. Label the plants with the numbers 1, 2, 3 and 4.
3. Give plant 1 one-half teaspoon of water every day.
4. Give plant 2 one teaspoon of water every day.
5. Give plant 3 one tablespoon of water every day.
6. Give plant 4 two tablespoons of water every day.

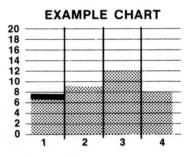

What happens:

1. Measure each plant daily for ten days.
2. Color in the chart for plant growth. Keep adding on to the colored column as the plant grows. If the plant dies, draw a black line across the top of the colored column.

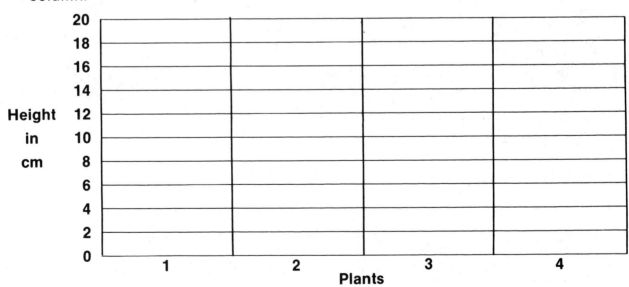

Puzzling out the results:

When your experiment is done, check the results. How did the amount of water affect plant growth?

A

Name _____

EXPERIMENT FOR YOUNG WIZARDS

Can plants find light?
Build this plant maze and find out.

Supplies needed:
   1 sprouting potato
   1 pot big enough for the potato and soil
   Stiff cardboard
   Masking tape
   Scissors
   Medium-size box with lid or flaps that can be closed tightly

How to do it:

1. Plant the potato in the pot so that the sprout is above the dirt.

2. Use the cardboard and tape to create short walls inside the box. Notice in the picture that the walls do not go all the way across the box. There must be room for the plant to get through.

3. Cut a hole in one side and place the potted potato inside the box at the opposite side.

4. Close the lid tightly.

5. Open the box each day only long enough to sprinkle water on the potato. Be careful not to overwater the potato. It will rot.

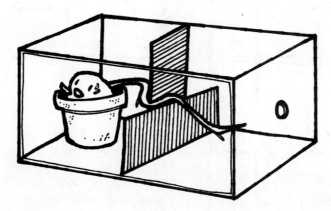

What happens:

1. How many days does it take the potato sprout to grow to the light?

_____

2. Why do you think that the potato sprout is so pale?

_____

Puzzling out the results:

    Plants do not have eyes to see — even though we call potato sprouts "eyes." How do you think the plant found its way to the sunlight?

Name _____

PUZZLE FUN

Can you help the plant find its
way through the maze to the sunlight?

I, TD   Name _____

HOW DOES WATER GET FROM THE GROUND TO A PLANT'S LEAVES AND FLOWERS?

Supplies needed:
- 2 tall glasses
- 2 colors of food coloring
- 1 long, fresh stalk of celery with leaves
- 2 pill bottles
- 1 long-stemmed white carnation (or other white flower)
- Knife or razor blade

How to do it:

1. Fill one tall glass half full of water and add two drops of food coloring.

2. Place the stalk of celery in the colored water and let sit for two hours.

3. Fill each pill bottle with water. Place them side by side inside the other tall glass. If you can't find a glass big enough, use a tall can.

4. Add four drops of two different colors of food coloring — two drops of one color to each pill bottle.

5. Carefully slit the stem of the flower up about half way.

6. Place the flower in the glass with half of its split stem in each of the pill bottles. Let the flower sit in the colored water for two hours.

7. Be sure that your plants are in a sunny place or under a light. Water rises much faster in a plant when the plant is in a light place. The tubes that carry water through a plant's stem are called *xylem* (zī'ləm) tubes.

What happens:

1. Color the flower and the celery stalk in the picture to show what happened.

2. Cut open the celery stalk. Can you see the tubes that carried the colored water to the leaves?

Puzzling out the results:

How did the water get to the flower and the leaves?

A

Name _____

EXPERIMENT FOR YOUNG WIZARDS

What kind of gas do plants give off?

Supplies needed:
- Large fish bowl or small fish tank
- Several shoots of water weed such as Anacharis
- 3 clip clothespins
- 1 plastic funnel
- 1 small clear glass pill bottle
- A friend
- Matches
- 1 wood toothpick

How to do it:

1. Fill the bowl or tank with water and place the plant in the water.

2. Clip the pins on the funnel and place over the plants as shown in the picture.

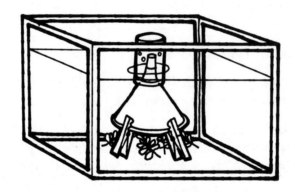

3. Fill the pill bottle with water and put the open end over the top of the funnel. Gas will slowly bubble up to push out the water. This process only happens when the plants are in light. Place your experiment in a sunny spot or under a light.

4. When the bottle is nearly filled with gas, remove it from the water. Put your thumb over the opening and keep the bottle upside down to keep the gas from rushing out.

5. Have a friend light the wood toothpick and blow it out. While it is still glowing, poke it into the bottle of gas.

> Two gases are important to plant and animal life: *oxygen* and *carbon dioxide.* If a glowing **stick** is put into oxygen, it will burst into flame. If a glowing **stick** is put into carbon dioxide, it will go out.

What happens:

1. What happens to the toothpick when you put it into the bottle?

2. What gas has the plant given off? _____

Copyright © 1980 — THE LEARNING WORKS

A

Name _____

Puzzling out the results:

Why do you think it is important for a pond or lake to have water plants?

Why do you think it is important for us to have trees and plants growing on our earth?

SECRET MESSAGES

Invisible ink was first used in 1776 during the Revolutionary War. Silas Deane, a member of the American **colonists'** Committee of Secret Correspondence, arrived in France with instructions to purchase military supplies for the colonies. These instructions had been written by John Jay in invisible ink. The invisible ink had been invented by Sir James Jay.

Some plant juices change color when they are heated. You can use this fact to send secret messages.

Lemon juice is one of the best juices to use. Write your message on plain white paper by dipping a paint brush or toothpick into the juice.

When the juice dries, there will be little or no trace of your message.

Send your secret message to a friend with these directions.

To read this message,
hold the paper close to a hot light bulb.
Be careful not to get burned.

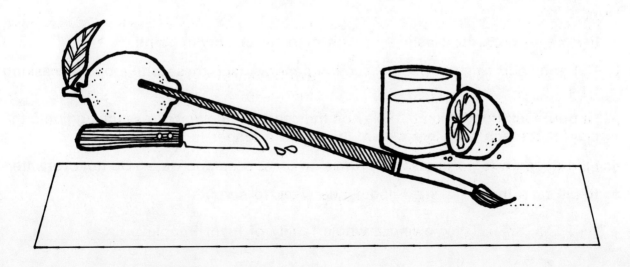

Copyright © 1980 — THE LEARNING WORKS

Name _____

FUN PAGE

How would you like to grow your own Plant Man complete with hair that you can trim into any style you want?

Follow these directions and your Plant Man will soon sprout into life.

Supplies needed:
- ½ plastic egg
- Permanent magic markers or acrylic paint and brush
- Construction paper
- Scissors
- Masking tape
- 1 quarter-sized lump of modeling clay
- Potting soil
- Grass seed

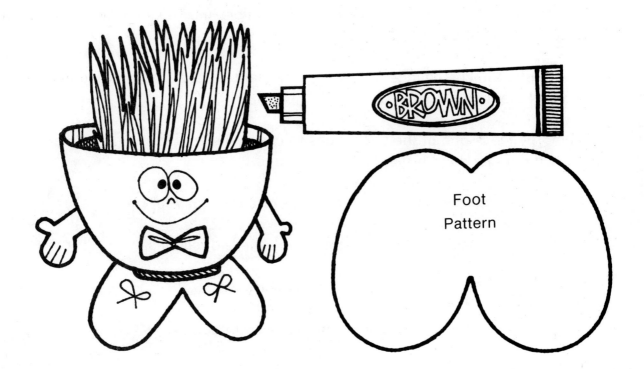

How to do it:

1. Paint a face on the plastic egg with markers or acrylic paint.

2. Cut arms and a tie out of construction paper. Attach these with loops of masking tape.

3. Cut out feet. Put the lump of clay on the center of the feet. Press the egg onto the clay. If the egg does not sit up, you need a bigger lump of clay.

4. Fill the egg with potting soil. Sprinkle on grass seed and water. Do not overwater.

5. It will take the grass seed about one week to sprout.

Try growing a whole family of Plant People.

What if Jack's mother had not thrown the magic beans out the window? What could Jack have done to make sure the seeds sprouted? How could Jack have helped his special beans to become the best beans possible?

GREEN FRIENDS

_____

Has successfully mastered GREEN FRIENDS

Date: _____

GETTING TO KNOW ME

BACKGROUND

Everything is made up of tiny, invisible particles called *atoms* and *molecules* — groups of atoms. Even though these particles are too small for us to see, they are big enough to trigger our senses.

The human body is an amazing structure able to sense tiny particles of chemicals (smell and taste), interpret vibrating waves or molecules (hear), and analyze reflected light (see). These plus the additional senses of touch and balance allow us to be aware of our environment and respond to it.

The nose (and nasal cavity) is the center for our sense of smell. Olfactory cells that line the nasal passages have tiny, hair-like endings. These specialized nerve cells are stimulated by molecules of chemicals in the air. As in any nerve, electrical impulses travel from the olfactory nerves to the brain.

The olfactory nerves quickly adapt to smells. Sniff some perfume and then go on sniffing. You will soon stop noticing the odor. The brain does, however, store a vast memory bank of smells. We can identify things and people by the way they smell. There is never any blending of smells. Even if molecules from roses and lilacs are breathed in at the same time, each smell is interpreted individually.

The ears, eyes, nose, and throat are all connected. Crying makes the nose run. A cold may cause the ears to seem plugged. A tube — called the *Eustachian tube* — connects the ears to the throat. Its main job is to equalize pressure in the middle ear. A change in pressure causes a feeling of pressure in our ears. Swallowing forces air up the Eustachian tube and equalizes the pressure.

Our sense of balance is controlled by the *cochlea* — an organ of the inner ear. The cochlea is a shell-shaped organ filled with fluid and many tiny nerve endings. As the fluid sloshes back and forth over these nerve endings, we have a feeling of how our head is positioned in relation to the rest of our body.

The sense of taste is limited to the tongue in the human body. Insects can taste through their feet and a number of fish can taste through the lateral line — a line of specialized cells running horizontally down their body. It's an interesting idea to imagine what it would be like if we could taste by touching our food.

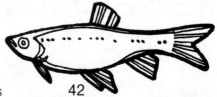

Unlike smell, many tastes blend together. One reason that we like sweet things so well is that this is the first taste that the tongue is able to sense as food enters the mouth.

Vision is our most used sense. The human eye is similar to a camera. The pupil is the opening through which light enters. The size of the pupil is controlled by a circular muscle — the *iris*. The light then passes through a jewel-like lens and a jelly-like fluid to reach the light-sensitive nerve cells of the *retina*. The retina has rods for black and white vision and cones for color vision.

Any time light passes through matter of different densities, it bends — refracts. In a normal eye the bending is enough to make the light focus on the retina. The image that focuses on the retina is upside down. It is up to the brain to interpret this image.

WORD BOX:

| symmetry | sensor | papillae |
| saliva | optical illusion | |

LEARNING OBJECTIVES:

1. Students will be able to identify bilateral symmetry.

2. Students will understand the main touch sensations and how they are spaced on the body.

3. Students will understand how smell and taste are related, be able to identify the four main tastes and understand how these tastes are sensed.

4. Students will understand how air vibrations become sounds.

PREDICTIONS:

1. Can you taste something if you can't smell it?

2. Show a picture of half of a person. What do you think the other half looks like? How do you know?

3. The main senses of the skin are heat, cold, touch, and pain. Which sense do you think your skin has the most sensors for?

DISPLAY AND BULLETIN BOARD IDEAS:

1. Divide a bulletin board into sections for each of the main senses — hearing, seeing, smelling, touching. Have a picture of the main sense organ in each section. Have at least one sample picture in each section of things that show something to do with that sense. Have students cut pictures out of magazines to add to the display. Also include small word cards with words that have to do with each sense. For instance, *soft* is a touch word, *bitter* is a taste word and *crunch* is a hearing word.

2. Put out a feel box (see the directions below). Change what is in it every day or every other day. Have a contest and let the students submit entries of what they thought was in the feel box each day. In case of duplicate answers, a drawing could be held for the winner of a small reward or special privilege.

FEEL BOX

1. Cut a hole in one side of a shoe box big enough for your students to slide their hand through.

2. Cut off the toe of an old sock and tape this open toe end inside the shoe box through the hole.

3. Put an object in the box such as a sponge, several golf tees or cotton balls. Put the lid on the box and tape it shut. Tell everyone that they are on their honor not to peek. The object is to be examined only by reaching through the sock to the inside of the box.

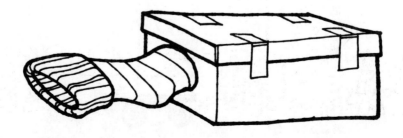

EXTRA SPARK STARTER:

1. Make a tape of a variety of different sounds such as an alarm clock, a doorbell, a car engine, a train, an appliance running, a nut cracking, a paper being crumpled. How many sounds can your students identify just by listening to them?

2. Run a rope around the room or in another room through a prepared course. Put away anything that could break and put out a variety of things to feel. Have the students close their eyes or tie on blindfolds of strips of old sheet.

   After the students have walked slowly through the course, ask them to write about everything that they experienced. Make a chart together. Talk about what they wrote down and how they decided what the objects were.

MATERIALS NEEDED:

Small mirror
Full-length mirror
Colored ball-point pen
Vinegar
Potato
White construction paper
Small box about 6 cm wide and less than 12 cm long (such as an individual serving cereal box or the bottom half of a ½ pint milk carton)

Scrap piece of 1 x 2 softwood
Scissors
Metric ruler
Hairpin
New, clean paintbrush
Salt
Onion
Black fine-line marking pen
Screw eyes
Spool of thread or a cork

Piece of 1 x 2 wood 40 cm long
Colored construction paper
Plastic cups
Sugar
Black coffee
Knife
Gallon-size plastic milk jug
Monofilament fishing line
Tacks or small nails
Hammer
Pencil

OPTIONAL MATERIALS:

Fingerpaint
Crisp cereal (Special K or Rice Krispies)

Margarine
Large spoon
9 x 13 cake pan
Crunchy peanut butter

Electric hot plate
Double boiler
Karo syrup
Chocolate chips

ADDITIONAL SOURCES: BOOKS

Johnston, Ted. Science Magic with Chemistry and Biology. Arco Publishing Company, Inc., 1975.

Podendorf, Illa. 101 Science Experiments. Children's Press, 1963.

Seymour, Simon. Kitchen Chemistry. Viking Press, 1971.

ADDITIONAL SOURCES: FILMSTRIPS

<u>Biology of the Human Body</u>. Intermediate level. Set of 8 filmstrips and sound. Encyclopedia Britannica Corporation.

<u>Slim Goodbody: Your Body, Health and Feelings</u>. Primary and intermediate level. Multi-media kit. SVE-Society For Visual Education, Inc.

<u>The Human Machine</u>. Grades 6-12. Set of 8 filmstrips and sound. Coronet.

<u>Your Senses and How They Help You</u>. Grades K-4. Set of 2 filmstrips and sound. National Geographic Society.

ANSWERS:

p. 49: Challenge Picture

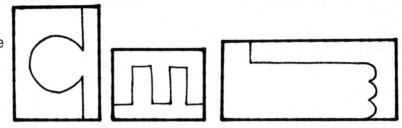

p. 50: What happens:
The number of dots for each sense will vary with the individual. In general, they are distributed in this order: pain, touch, cold, and heat.

Puzzling:
There are more pain receptors.
This is important because it warns the person of danger.
Heat has the fewest sensors in the skin.

p. 51: On your forearm, you will feel two points when the points of the hairpin are far apart and one point when the points are close together. You will feel two points during both tests on your fingertips.

p. 52: You will not be able to taste the sugar until saliva begins to moisten your mouth and tongue.
Taste areas of the tongue:

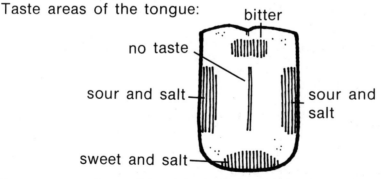

p. 53: Your sense of smell is closely linked to your sense of taste. Even though you have potato on your tongue you will have an onion taste in your mouth. Onion atoms of smell are so strong a chemical that just smelling them gives you the sensation of tasting onion.

p. 54: 1. You see a floating piece of finger.
You see this because the eye is not able to focus at distant and close objects at the same time.

EVALUATION:

Each of these statements is false in some way. Cross out what is false and print above it the word or words that would make the statement true.

1. In bilateral symmetry, the right side of something is completely different from *(nearly the same as)* the left side.

2. Our touch sensors are farthest apart on our fingertips. *(closest together)*

3. There are more sensors for touch than any other sense in our skin. *(pain)*

4. Glands in the mouth produce a liquid called papillae to moisten food. *(saliva)*

5. The center of the tongue senses bitter. (or the back of the tongue senses bitter) *(has no taste)*

6. The tip of the tongue senses sour and salt. *(sweet)*

7. The sense of hearing and tasting are closely related. *(smelling)*

8. In the eye, the special sensory nerves called cones are for black and white vision. (or cones are for color vision) *(rods)*

9. The spot where the olfactory nerve enters the eye is called the blind spot. *(optic)*

10. Sound is really vibrating waves of light. *(air)*

EXTENDED LEARNING:

1. Make a collection of texture rubbings. These can be in a labeled booklet or in a collage.

2. Blindfold students and have them do a clay sculpture.

3. Write a radio play complete with sound effects. Record the play on a tape recorder so that everyone can listen to the results.

4. Make other musical instruments and start a homemade band.

5. Make a flip book to experiment with animation and seeing movement.

6. Have a visit from a doctor, an optometrist, or a paramedic.

7. Write a story describing the action only in terms of one sense (the writer can only see what happened or only hear what happened).

8. Describe something using only one sense (describe a pencil only by how it feels). Remember the poem about the blind men and the elephant?

Copyright © 1980 — THE LEARNING WORKS

THE HOMEMADE MAN

The lightning streaked yellow across the black sky. The huge stone house was dark except for a dim light that glowed from the windows under the roof.

Inside his attic workroom, Dr. Frankenstein was bent over a long table. Coils of wire connected the table to weirdly shaped machines.

"Bring me the right arm," Dr. Frankenstein's voice boomed with the thunder.

Igor, the doctor's chubby, bald-headed assistant, waddled into the room carrying a big metal tray. A single, muscular, human arm was draped across the tray.

With the needle and thread, Dr. Frankenstein sewed the arm to the body on the table.

"It is done." Dr. Frankenstein smiled as he wiped his hands on his lab coat.

Quickly, Dr. Frankenstein attached a wire leading from the table to the windows. Then he and Igor moved back from the table.

The next lightning bolt shot through the window, smashing the glass. The tremendous charge crackled through Dr. Frankenstein's machines and quivered over the body on the long table.

Slowly, a giant man sat up. The man's hair stood up like a bristly brush. The lightning's glow still sizzled around him.

"He moves," Dr. Frankenstein said to Igor, "but is my monster really alive?"

Dr. Frankenstein wanted his monster to be like a real human being. You are a real human being. How do you experience your world?

In the following experiments, you will investigate your body and your senses.

Name _____

ARE BOTH HALVES OF YOU EXACTLY THE SAME?

Supplies needed: Construction paper
                   Centimeter ruler
                   A full-length mirror (if possible)
                   A small square or rectangular mirror

How to do it:

1. Look at your right hand and your left hand. Can you see any differences between the two hands? _____

2. On a piece of construction paper, draw around your left and right hands.

3. Measure each of your fingers from your first knuckle bump to the tip of your finger (soft tip, not the end of the nail).

4. List the measurements on the drawing of your hands.

5. Look in the full-length mirror. In what ways is your right side like your left side?

In what ways is your right side different from your left side?

Symmetry is the way body parts are arranged. The most common kinds of symmetry found in living things are *radial symmetry* and *bilateral symmetry*. A starfish has radial symmetry. All parts of the starfish's body are arranged around a central part. The human body has bilateral symmetry. The human body has two nearly equal sides.

CHALLENGE PICTURE

What would the other half of each picture look like? Place a small mirror on edge against the unfinished side of each picture.

Can you match the pictures to their other halves?

Copyright © 1980 — THE LEARNING WORKS

Name _____

WHAT SENSES DO YOU HAVE IN YOUR SKIN?

Supplies needed: A colored ball-point pen or colored pencil
 A friend

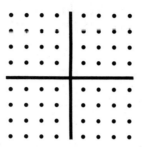

How to do it:

1. Draw a large cross on the back of your hand.

2. Make four rows of four dots each to fill in each section. Make the dots close together.

3. Use a pencil with a rounded point. Have a friend touch each dot one at a time. Really think about what you feel.

What happens:

1. How many dots give you a sense of touch? _____

2. How many dots give you a feeling that the pencil point is cold? _____

3. How many dots give you a feeling that the pencil point is warm? _____

4. How many dots give you a feeling that the pencil point is sharp? _____

5. How many dots give you a feeling of pain (even though the pencil is not pressed hard)? _____

Puzzling out the results:

   Sensors are nerves that receive signals and respond to them. In the skin, your body has sensors for *touch, warmth, cold, sharpness,* and *pain.*

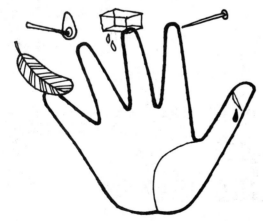

Which of these senses has the most sensors in your skin? _____

Why do you think that it is important for this sense to have the most sensors?
_____
_____

Which of these senses has the fewest sensors in your skin? _____

Name _____

HOW CLOSE TOGETHER ARE YOUR TOUCH SENSORS?

Supplies needed: A hairpin

How to do it:

1. Spread a hairpin's points about four cm apart.

2. Touch your forearm and then touch your fingertip.

3. How many points do you feel on your arm? _____
   On your fingertip? _____

4. Now push the hairpin's points as close together as possible.

5. Touch your forearm and then touch your fingertip.

6. How many points do you feel on your arm? _____
   On your fingertip? _____

Your touch sensors are more than one cm apart on most parts of your body, but on parts meant for touching — such as your fingertips — the touch sensors are very close together.

FUN SPOT

Did you know that nobody has fingerprints like yours? No two of your fingers have prints that are alike either. You can use your "special" fingerprints to make print animals.

Supplies needed: Several colors of finger paint
White construction paper
Black fine-line marking pen

How to do it:

1. Be sure that the paint is neither too thin nor too thick.

2. Start with your hands clean and dry.

3. Spread a small amount of paint on a folded paper towel or scrap of paper.

4. Press your thumb into the paint and then press it onto the white paper.

5. Make fingerprints and combinations of thumb and fingerprints.

6. After the prints have dried, decide what animals you would like to make these prints become. Use the marking pen to add details.

Name _____

HOW DO YOU TASTE?

You know that your tongue is the part of your body that allows you to taste. Look at your tongue in the mirror. Your tongue is covered with bumps called *papillae.* Inside each papilla are special sensory nerves called *taste buds.* These taste buds allow you to sense *4* main tastes: *sweet, sour, salt* and *bitter.*

Can you taste if your tongue is dry? Try it and see. Dry your tongue with a clean washcloth. Quickly, put a little bit of sugar on your tongue. You will be able to feel the sugar, but can you taste it? _____

What happened to your tongue and mouth before you were able to taste the sugar? _____

Glands in the mouth produce a liquid — *saliva* — that floods over the food, carrying tiny particles into the taste buds.

WHERE DO YOU TASTE SWEET, SOUR, SALT AND BITTER?

Supplies needed:
- 1 new, clean paintbrush
- 1 plastic cup with ¼ cup of black coffee
- 1 plastic cup with ¼ cup of vinegar
- 1 plastic cup with ¼ cup of very salty water
- 1 plastic cup with ¼ cup of very sugary water
- 1 cup of plain water

How to do it:

1. Dip the paint brush into the black coffee.

2. Carefully, touch the brush to the back, sides and front of your tongue.

3. Rinse the brush well. Rinse your tongue by swishing water in your mouth.

4. Repeat with each of the other liquids.

What happens:

1. Where did you best taste the bitter coffee? _____

2. Where did you best taste the sour vinegar? _____

3. Where did you best taste the salty water? _____

4. Where did you best taste the sugary water? _____

The center groove of your tongue has no taste buds and no sense of taste.

YOUR SENSE OF SMELL

Your nose is your body's smell center. Inside your nose and nasal cavity, there are many cells called *olfactory nerves*. Each cell has special hair-like endings to pick up chemicals in the air and send messages to the brain.

The olfactory nerve hairs are not the hairs that you can see in your nose. The hairs that you can see keep out dirt, as does the mucus (watery liquid).

olfactory nerves

DOES YOUR SENSE OF SMELL MAKE A DIFFERENCE IN HOW YOUR FOOD TASTES?

Supplies needed: 1 piece of potato   1 piece of onion

How to do it:

1. Pinch your nose.
2. Put the piece of potato on your tongue.
3. Unpinch your nose and hold the piece of onion under your nose.

What happens:

1. You know that you have a piece of potato on your tongue. Does it taste like potato or onion? _____
2. Take the potato out of your mouth and take a deep sniff of the onion.
3. Can you still taste the onion even without any food in your mouth? _____

FUN SPOT

Here's a no-bake cookie recipe to tempt your tongue and delight your nose. You may need an adult's help.

Supplies needed:
- Electric hot plate
- Double boiler
- 1 cup sugar
- 1 cup white Karo syrup
- 2 tablespoons margarine
- Large spoon
- 1½ cups crunchy peanut butter
- 6 cups Special K cereal or or Rice Krispies
- 9 x 13 cake pan
- 1 6-ounce package of chocolate chips

How to do it:

1. Put the sugar and Karo syrup in the double boiler. Be sure to have water in the bottom part of the double boiler. You can use two pans that fit together loosely if you do not have a double boiler. Add margarine.
2. **Stir the mixture on medium heat until it comes to a rolling boil. Remove from heat.**
3. Stir in the peanut butter and cereal. Press into greased cake pan.
4. Put the chocolate chips in the double boiler and stir just until melted.
5. Spread the melted chocolate over the cookies. Let cool and then cut and eat.

Name _____

HOW DO YOU SEE?

The sense that we use the most is our vision. On the inside back of the eye is an area called the *retina*. The retina has special sensory nerves called *rods* for black and white vision and *cones* for color vision.

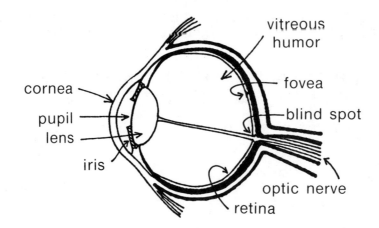

You see when these nerves send messages along the optic nerve to your brain. However, you can't see with all of your retina. Where the optic nerve enters the eye, there is a blind spot. Just above the blind spot is the *fovea*. The fovea has a lot of cones and is the spot of best vision in your eye.

The lens is moved by muscles so you can see things clearly at a distance or close up. When you try to look both at something far away and something close to your eyes at the same time, your eyes are fooled.

Hold the tips of your forefingers together a little less than arm's length from your face. Look just above your fingers at something in the distance. Stare at that spot while you slowly move your fingers toward your face.

SURPRISE!

What do you see between your two fingertips? _____

Why do you think you see this? _____

_____

Name _____

FUN PAGE

When you trick your eyes, the tricks are called *optical illusions*. Try these optical illusions just for fun.

Big Circle—Little Circle

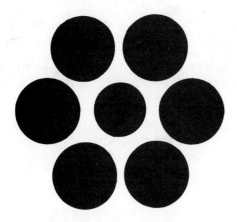 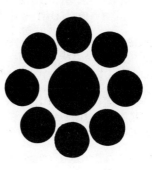

Which center circle is bigger? Measure them to find out. Surprise!

Your eye not only compares the two middle circles to each other, it also compares the center circles to the other circles around them.

The Crooked Square

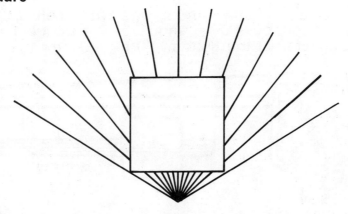

Is this square really crooked or do the lines only make it look uneven? Measure the sides and find out.

Copyright © 1980 — THE LEARNING WORKS

A

Name _____

EXPERIMENT FOR YOUNG WIZARDS

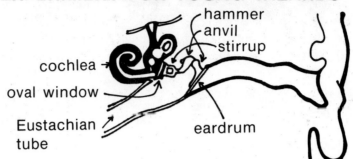

Your ear is an amazing sound receiver. The outer ear funnels sound to the *eardrum.* The eardrum is a thin membrane that vibrates like a tightly stretched balloon.

Three tiny bones pass these vibrations on to another thin membrane — the *oval window.* On the other side of the oval window, the vibrations tickle tiny nerve endings, sending electrical impulses to your brain.

Any sound is really waves of air molecules. The speed at which these molecules vibrate determines what sound you hear. Scientists have found that we hear a sound that we call middle C when the air vibrates 256 times a second.

You can make musical instruments to investigate sound yourself.

MILK JUG BANJO

Supplies needed:
- Empty plastic milk jug (gallon size)
- Scissors
- Piece of soft wood about 8 cm wide, 2 cm thick and 1 m long
- 6 screw eyes or hooks
- 4 m of monofilament fish line
- A spool of thread or a cork

How to do it:

1. Use sharp scissors to cut an H-shaped slot about two centimeters from the bottom of each side of the milk jug.

2. Slide the piece of wood through the slot.

3. Now you need to be able to attach the strings. With a pencil, mark where you will insert the screw eyes or hooks. They should be in a straight line across the top of the banjo and staggered at the bottom. Turn the screw eyes or hooks into the wood.

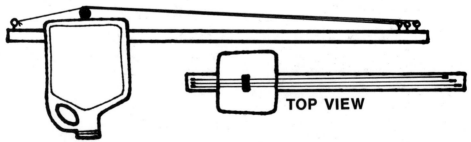

4. Cut a piece of monofilament fish line a little longer than the piece of wood for each string. Be sure it is tied tightly to the hooks.

5. Slide the spool or cork under the strings on the milk jug. This forms the bridge to lift the strings off the milk jug and allow the air under them to vibrate.

A   Name _____

ONE-STRING GUITAR

Supplies needed:
- A piece of 1 x 2 wood 40 cm long
- 2 tacks or small nails
- Hammer
- A piece of monofilament fish line a little longer than the wood
- Metric ruler
- Pencil
- Cardboard box about 6 cm wide and not longer than 12 cm (such as an individual serving cereal box or the bottom half of a ½ pint milk carton)

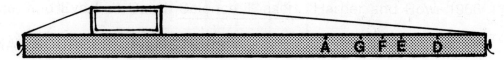

How to do it:

1. Hammer one tack or small nail into the top of the wood at each end. Do not hammer it all the way into the wood. You will need room to tie the fish line to the tack or nail.

2. Tie the fish line to each tack. Tie it very tightly.

3. Measure twenty-eight centimeters from one end. Mark this with the pencil.

4. Slide the box under the string so that the front edge is on the line.

5. For your musical scale:
   Measure four centimeters from the end farthest from the box (front edge). Mark this spot. This is D.
   Measure seven and a half centimeters from the top. This is E.
   Measure nine centimeters from the top. This is F.
   Measure eleven centimeters from the top. This is G.
   Mark fourteen centimeters from the top. This is A.
   To play each of the notes that you marked, press the string against the wood with the side of a pencil on the note's mark while you pluck the string near the box.
   To play middle C, pluck the string without pressing it down.

   Follow these musical patterns to play each song.

MARY HAD A LITTLE LAMB

E D C D E E E   D D D   E G G
E D C D E E E E D D E D C

TWINKLE, TWINKLE, LITTLE STAR

C C G G A A G F F E E D D C
G G F F E E D G G F F E E D
C C G G A A G F F E E D D C

Name _____

FUN PAGE

How many words about your senses
can you find hidden on Dr. Frankenstein's monster?
Use the Clue Box to help you.

SMELL	SIGHT	EYE	EAR
NOSE	COCHLEA	FOVEA	BLIND SPOT
BRAIN	TOUCH	TASTE	COORDINATION
TONGUE	SWEET	SOUR	BITTER
SALT	BALANCE	PAPILLAE	RETINA

Copyright © 1980 — THE LEARNING WORKS

How should Dr. Frankenstein's monster be able to experience his world?

Can you list all of your senses and tell how each helps you know more about the world around you?

GETTING TO KNOW ME

_____

Has successfully mastered GETTING TO KNOW ME

Date: _____

THE MAGIC OF MAGNETS

BACKGROUND

People have been fascinated by magnets since prehistoric time. To early man, the natural magnets — lodestones — seemed magical.

Today, we understand that whenever the tiny particles that make up matter — molecules and atoms — are made to line up in orderly rows, the object becomes a magnet.

A number of different kinds of matter will become at least slightly magnetic, but iron and steel become the strongest magnets and keep their magnetic powers longest. Lodestones are a type of iron ore.

Magnets can be created by rubbing the object in one direction only, with one pole of a magnet. Magnets can also be created by passing an electric current through an object in one direction only. In fact, an electric current has a magnetic force field around itself.

Magnets are made in bar and horseshoe shapes. Horseshoe magnets are generally stronger than bar magnets.

A magnet can be destroyed by a sudden hard blow, by dropping it, or by running an electric current through it in two directions. High heat will also destroy a magnet. In other words, anything that disrupts the orderly pattern of the molecules or atoms alters the object's magnetic powers.

What are a magnet's powers? A magnet will attract — pull toward itself — solid objects made of iron or steel.

Because of the way molecules are arranged, magnets are said to have two poles — a north and a south pole. If two magnets of equal strength were brought close to each other, the two north poles would repel each other, or give you a feeling that they were pushing apart. The two south poles would do the same. The north and south poles would attract, or give you a feeling that they were pulling together.

A magnet's power extends beyond itself in an area called its *force field*. This force field can be seen when the magnet is placed under a piece of paper and the paper is sprinkled with iron filings. The filings line up along the lines of force.

These lines of force will also extend through all other forms of matter. A magnet can attract a steel paperclip through your hand, a piece of glass, even a rock. The magnet's power is stopped only if the object is thicker than the width of the force field.

The earth is a giant magnet. The earth's magnetic poles are not exactly the same as true North and South. The magnetic North Pole is about 1,400 miles south of true North. Magnetic North is close to Bathurst Island off the northern coast of Canada.

Any time a magnet is suspended in space, its south pole moves to point toward the earth's magnetic north pole. This creates what we call a *compass*.

WORD BOX:

| magnetism | magnetic | force field |
| compass | | |

LEARNING OBJECTIVES:

1. Students will be able to identify the kinds of things that a magnet will attract.

2. Students will understand how magnets interact.

3. Students will understand what a magnet's force field is like.

4. Students will be able to make magnets with metal and create an electromagnet by using electricity.

5. Students will understand how a compass works.

PREDICTIONS:

1. How will magnets react with each other?

2. How could you make a magnet?

3. What will block a magnet's force field?

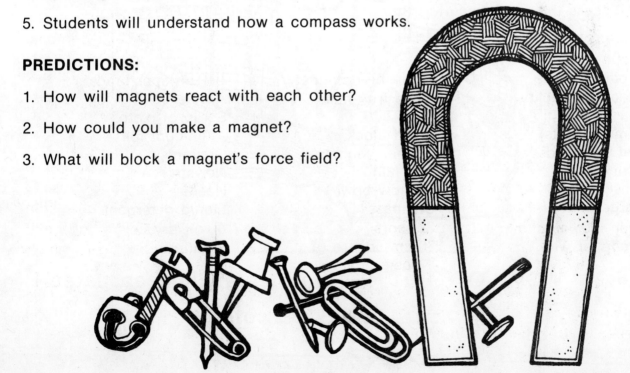

Copyright © 1980 — THE LEARNING WORKS

DISPLAY AND BULLETIN BOARD IDEAS:

1. Make a large horseshoe magnet out of posterboard or, for a three-dimensional effect, out of styrofoam — spray-painted a bright color. Put the magnet either in the center of the bulletin board or in the center above the bulletin board. Then add pictures of a variety of different objects. Again, for a three-dimensional effect, you could put some actual objects among the pictures. To challenge thinking, add the question, "Which objects will the magnet attract?"

2. Use the compass code on p. 72. Each day put up a new coded message.

3. Label a display area MAGNETS IN ACTION. Have pictures of things that use magnets. Encourage your students to add more examples. Remember that anything that uses an electric motor uses magnets. Magnets are one of the parts of the motor.

EXTRA SPARK STARTER:

1. Divide the class into small groups. Equip each group with a magnet. Challenge each group to see how many things they can find in the room that the magnet will attract. The group must keep a list. Set a time limit.

   Make a chart together of the results. The team with the most correct objects wins. Be prepared for multiple winners.

   Discuss as a class what the objects that were attracted by the magnet were made of. Draw a general conclusion about metals, mainly iron and steel, being attracted to magnets.

   Make a chart together of things that are not attracted by a magnet. Leave the charts on display for everyone to think about and refer to as they do the experiments.

2. Discuss the predictions together.

MATERIALS NEEDED:

2 bar magnets	Horseshoe magnet	Brass paper brads
Steel paper clips	Large iron nails	A glass
Plastic toy (no metal parts)	Paper	Scrap of cloth
Rubber bands	Wood block	Aluminum foil
String	Copper pennies	Crayons
Masking tape	Iron filings	Newspaper
Large cork	Shallow bowl	Needle
Thin insulated copper wire	Compass	Liquid detergent
	Scissors	6 volt dry cell battery
	30 cm square of posterboard	

ADDITIONAL SOURCES: BOOKS

Gibson, Walter B. Magic With Science. Grosset & Dunlap, 1968.

Kjellstrom, Vjorn. Be Expert With Map and Compass, The Orienteering Handbook. Charles Scribner and Sons, 1976.

Podendorf, Illa. 101 Science Experiments. Children's Press, 1963.

Vivian, Charles. Science Experiments & Amusements For Children. Dover Publications, 1963.

Wyler, Rose. What Happens If . . . ? Walker and Co., 1974.

Wyler, Rose, and Eva Lee Baird. Science Teasers. Harper and Row, 1966.

ADDITIONAL SOURCES: FILMSTRIPS

Making Things Work. Primary and intermediate level. Filmstrip and sound. Encyclopedia Britannica Corporation.

Science and Imagination. Grades 1-5. Multimedia kit. Coronet.

ANSWERS:

p. 67: Lists will vary. All the objects attracted to the magnet will be solid objects, metallic and mostly made of iron or steel.
   The two kinds of things that magnets will attract best are iron and steel.

p. 68:

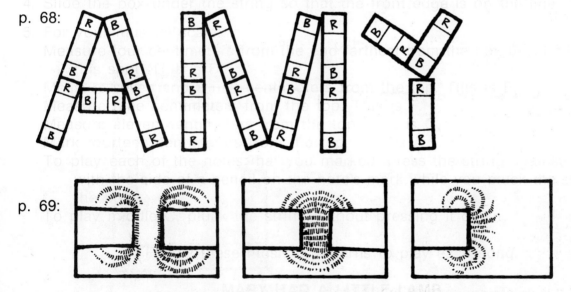

p. 69:

Puzzling: The first is like A and the second is like B.

p. 70: The results will vary with the strength of the magnet. In general, the strength should increase with more strokes and the stronger magnet should retain its power longer.

p. 71: The answers will vary. The magnet will attract through everything that isn't wider than the width of its force field.

p. 72: What happens: The compass and the needle point north.

   The compass's coded message says: "This cat just saw a dog."

Copyright © 1980 — THE LEARNING WORKS

p. 74: What happens:
  2. The number of clips depends on how strong the electric current is.
  3. Without the current, the wire will not attract any clips.
  You know an electric current has a magnetic field because the charged wire and nail attracted like a magnet.

Challenge Picture

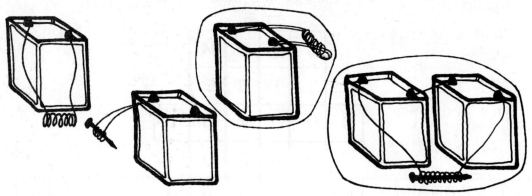

EVALUATION:

Circle everything on the list that a magnet would attract.

(Iron nail)   Window glass   Wood ruler
Aluminum pan   Plastic ruler   Rubber ball
(Steel paper clip)   Notebook paper   (Metal coat hanger)
Cotton shirt

Complete this statment:

_____(Iron)_____ and _____(steel)_____ make the best magnets.

Label the magnets N for north pole and S for south pole to show how they must be arranged to form this pattern.

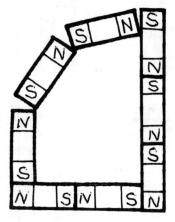

Tell why the electromagnet in this picture will not work.
  (Wire is not connected to both terminals of the battery.)

Tell two ways that it could be made stronger after you got it to work.
  (Add more coils.)
  (Add an iron nail core.)

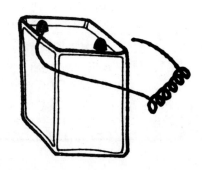

EXTENDED LEARNING:

1. Start an Adventure File with these ideas. Encourage students to try these experiments and to add other Adventure File ideas on their own.

    1. Will a magnet attract objects under water?
    2. How does a magnet affect a compass?
    3. How does combining magnets affect their power?
    4. How far can a magnet attract?
    5. Can you make a small electric motor using a horseshoe magnet?

2. Put out graph paper and invite students to create compass pictures for each other to work out. Make up another design yourself for your students to do. Display the finished designs.

3. Find out more about Marco Polo's adventures.

4. Find out about Samuel Morse.

5. Design a machine using magnets. Write a description of your picture, telling how it works. The wilder the better. Name the machine.

6. Find out about early explorers who depended on a compass for navigation.

7. Make up a story about being lost by yourself with only a compass to help you find your way home.

8. Find out about orienteering (navigating on land with a compass). Set up orienteering trails and courses to follow.

Copyright © 1980 — THE LEARNING WORKS

THE SPECIAL GIFT

In the thirteenth century, Marco Polo went to China. He found many wonderful things to take home. With the help of the emperor of China, he put together a whole caravan of treasures.

When Marco Polo was ready to leave, the emperor gave him one last special gift — a compass.

A small, flat, metal fish floated in a carved ivory bowl full of water. No matter which direction the bowl was turned, the fish always pointed in the same direction — North.

Marco Polo's eyes opened wide. "This is wonderful!" he said.

The emperor smiled.

Marco Polo looked closely at the floating metal fish. "What is the power of this metal? Is it magic?" he wondered.

The compass fish was a magnet. Was the magnet's power magic?

In the following experiments, you will investigate magnetism. Everything in nature is made up of tiny particles called atoms and molecules. In some things these particles can be made to line up in an orderly pattern. When the molecules are lined up, the object takes on special properties. It becomes magnetic.

As you experiment, you will be exploring the special properties of magnets and discovering how magnets are created.

WHAT WILL A MAGNET ATTRACT?

Supplies needed:
- Bar magnet
- Brass paper brad
- A glass
- Paper
- A wood block
- A rubber band
- A copper penny
- Crayons
- Steel paper clips
- Iron nail
- A plastic toy without metal parts
- A piece of any kind of cloth
- Aluminum foil

How to do it:
  Touch the bar magnet to each object one at a time.

What happens:
1. Fill in the chart as you discover which things will be attracted to a magnet and which will not.
2. Try any other kinds of objects (think what they are made of) and add these things to your list.

A MAGNET WILL ATTRACT	A MAGNET WILL NOT ATTRACT

Puzzling out the results:
What are the two kinds of things that magnets will attract best? _____

FUN SPOT

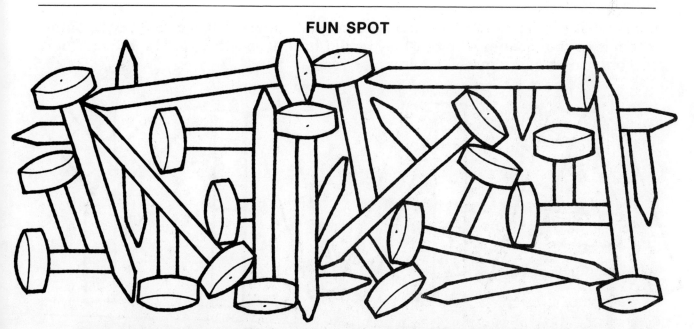

There are ten iron nails in this picture. Find them and color them red.

TD   Name _____

HOW DO MAGNETS REACT TO EACH OTHER?

Supplies needed: 2 large bar magnets   2 pieces of string 30 cm long
                   2 rubber bands   Crayons or colored pencils

How to do it:

1. Loop one rubber band over each end of the magnet so that it fits snugly.

2. Using one piece of string, tie one end to each of the rubber bands.

3. Tie the other string to the middle of the piece of string holding the magnet. Be careful not to drop the magnet. A sudden blow can destroy a magnet.

4. Suspend the magnet so that it can swing freely.

5. Hold the other magnet so that its south pole is close to the south pole of the suspended magnet.

6. Then hold the other magnet so that its north pole is close to the north pole of the suspended magnet.

7. Then hold the north pole of the other magnet so that it is close to the south pole of the suspended magnet.

What happens:

Using what you have learned about how magnets interact, color in the ends of the magnets in the picture below as they would have to be to form the letters. Color all north poles *blue* and all south poles *red.*

Can you make a magnet pattern for your name?

Name _____

WHAT IS A MAGNET'S FORCE FIELD LIKE?

Supplies needed: Newspaper   A 30 cm square of posterboard
                   2 large bar magnets   Iron filings

Marco Polo was amazed by the way the magnet could attract other metals without even touching them. How could the magnet do that?

The magnet's attracting power extends beyond itself. The area of magnetic attraction around the magnet is called its *force field*. Marco Polo couldn't see the magnet's invisible force field, but you can.

How to do it:

1. Spread out the sheet of newspaper.

2. Place the two magnets on the paper about two cm apart with two *like* poles facing each other.

3. Put the square of posterboard over the magnets and sprinkle on the iron filings. Sprinkle the filings evenly over the posterboard.

   Since iron is attracted to a magnet, the filings line up along the magnet's invisible lines of force.

4. Draw a picture in the box labeled "2 Like Poles" showing the pattern of the filings.

5. Carefully, pour the filings back into a jar.

6. Repeat, placing the two magnets with two *unlike* poles facing each other.

7. After you have drawn a picture of this pattern, repeat the experiment using only one magnet.

2 LIKE POLES   **2 UNLIKE POLES**   **1 MAGNET**

Puzzling out the results:

Using what you have learned about a magnet's invisible force field, draw a line to match the force field pictures to the pictures showing how the magnets might be arranged to cause these patterns.

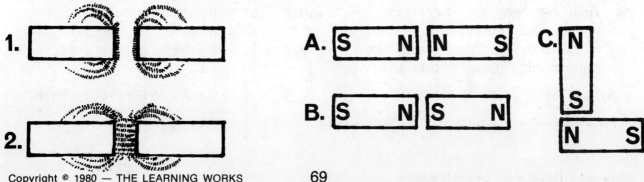

I, TD Name _____

HOW CAN YOU MAKE A MAGNET?

Supplies needed: 2 large iron nails   1 strong bar or horseshoe
                   12 steel paper clips   magnet

How to do it:
1. Stroke each iron nail with the magnet.
2. Stroke the nails in one direction only. Swing your arm out as you stroke so that the magnet is away from the nail on the return motion.

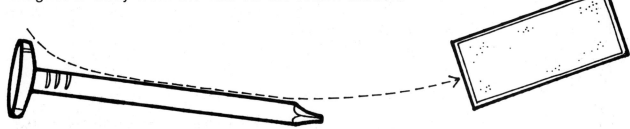

3. Stroke one nail twenty times. Stroke the other nail one hundred times.

What happens:
1. Hold the nail that was stroked twenty times over the pile of twelve clips. How many paper clips will the nail pick up as you bring it straight down on the clips? (Do not circle the nail over the pile.)

2. Hold the nail that was stroked one hundred times over the pile of clips. How many paper clips will the nail pick up as you bring it straight down on the clips?

Puzzling out the results:
   How do you think you could make the iron nail magnet stronger?

How long does the magnetic effect last? Repeat the test each day for three days and compare the results.

Day	Nail with 20 Strokes	Nail with 100 Strokes

What happens if you drop the nail magnet? Restrengthen the nails if you need to by stroking them with the magnet. The nails should be able to pick up as many clips as they originally did. Then drop the nails. How many clips are the nails able to pick up after being dropped?

NAIL WITH 20 STROKES   NAIL WITH 100 STROKES
_____   _____

Name _____

WHAT WILL A MAGNET ATTRACT THROUGH?

Supplies needed: Piece of string 12 cm long   4 or 5 books
                     1 steel paper clip   Samples of paper, wood,
                     Masking tape   glass, plastic, cloth,
                     A strong horseshoe   aluminum foil, rubber
                       magnet

How to do it:

1. Tie the string to the paper clip.

2. Tape the string to the table.

3. Put the horseshoe magnet on top of enough books to be just above but not touching the clip (the magnet's force field will pull the clip up but the string will prevent the clip from reaching the magnet).

4. Slide each of the objects between the magnet and the clip one at a time.

5. If the magnet will attract through the object, the clip will remain upright. If the object blocks the magnet's force field, the clip will fall.

What happens:

1. Fill out the chart as you discover which things will block the magnet's force field and which a magnet will attract through.

2. Try any other things you can think of (think what they are made of) and add these to your list.

A Magnet Will Attract Through	A Magnet Will Not Attract Through

TD   Name _____

HOW DOES A COMPASS WORK?

Supplies needed: A bar magnet   A shallow glass or plastic bowl
                   A needle   Liquid detergent
                   A large cork   A compass

The Earth is a giant magnet with a North and South Pole. The Earth's magnetic poles are not exactly the same as the Earth's true North and South Poles. The magnetic North Pole is about 1,400 miles south of the true North Pole. Magnetic North is close to Bathurst Island off the northern coast of Canada. See if you can find this island on a map. A compass points to magnetic North.

How to do it:
1. Magnetize the needle by stroking it in one direction only with one pole of the bar magnet. This will require forty or more strokes.
2. Push the needle through the middle of the cork.
3. Fill the bowl with water, add three drops of liquid detergent and stir well. The detergent keeps the cork from drifting to the edge of the bowl.
4. Place the cork-needle in the center of the bowl of water.

What happens:
Does the cork and needle turn in the water? (If not, you need to remagnetize the needle.) Using the real compass, what direction does the needle in the cork point?

_____

(Don't bring the real compass too close to the cork-needle; the two magnets will affect each other.)

✳✳✳✳✳✳✳✳✳✳✳✳✳✳✳✳✳✳✳✳✳✳✳✳✳✳✳✳

FUN SPOT

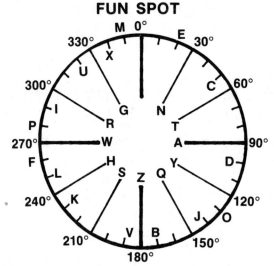

Use this special compass above to decode this secret message about the Fun Page.

| 60° | 240° | 290° | 210° | | 50° | 90° | 60° | | 140° | 320° | 210° | 60° |

| 210° | 90° | 270° | | 90° | | 100° | 130° | 330°. |

Can you write coded messages of your own?

Name _____

FUN PAGE

Plot an imaginary course using the compass directions below the graph. Fill each box with an X as you move through it.

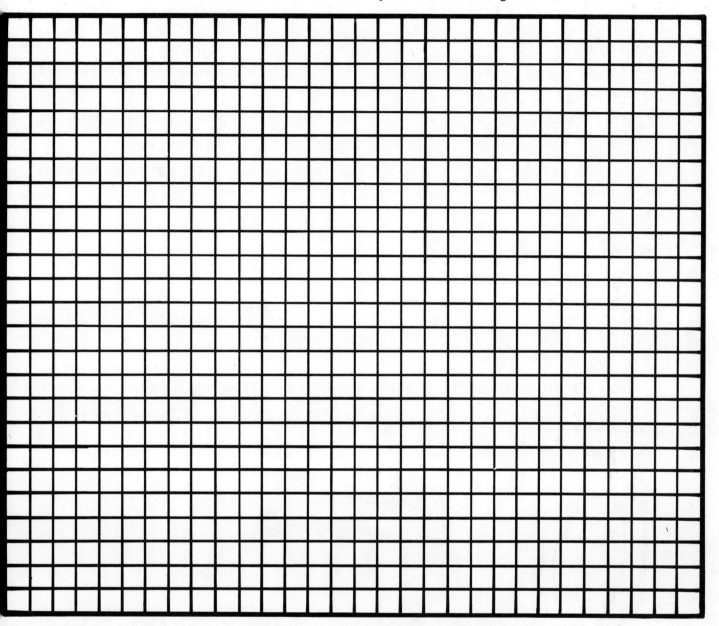

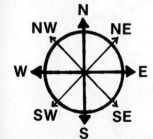

COMPASS DIRECTIONS
Start 10 down and 7 across from upper right-hand corner.
Do not count your starting square as number 1.

a. 5-N	b. 2-NE	c. 1-N	d. 3-SW	e. 4-S	f. 1-W	g. 1-S
h. 4-W	i. 2-SW	j. 3-W	k. 1-NW	l. 1-N	m. 1-SW	n. 1-S
o. 1-W	p. 1-SW	q. 2-S	r. 1-E	s. 2-SE	t. 7-S	u. 1-E
v. 2-N	w. 2-NE	x. 7-E	y. 4-S	z. 1-E	aa. 3-N	bb. 1-NE
cc. 1-N	dd. 1-NE	ee. 6-N				

Copyright © 1980 — THE LEARNING WORKS   73

Name _____

DOES AN ELECTRIC CURRENT HAVE A MAGNETIC FIELD?

Supplies needed: 3 meters of thin insulated copper wire
Scissors
Iron nail
1 6-volt battery
36 paper clips

How to do it:

1. Use the scissors to scrape about one cm of insulation from each end of the copper wire.
2. Wrap the wire fifty to one hundred times around the iron nail, forming a tight coil.
3. Fasten one end of the copper wire to each of the battery terminals.

   Be sure to disconnect the wire if it begins to feel warm. Do not leave the wires connected to the battery too long at one time — it weakens the battery.

What happens:

1. Bring the wire-wrapped nail close to the pile of paper clips.
2. How many clips can the nail pick up at one time? _____
3. Disconnect the wire from one terminal. How many clips will the coil and nail pick up this time? _____

Puzzling out the results:

How do you know that an electric current has a magnetic field?

CHALLENGE PICTURE

Circle the pictures that show ways to make an electromagnet stronger.

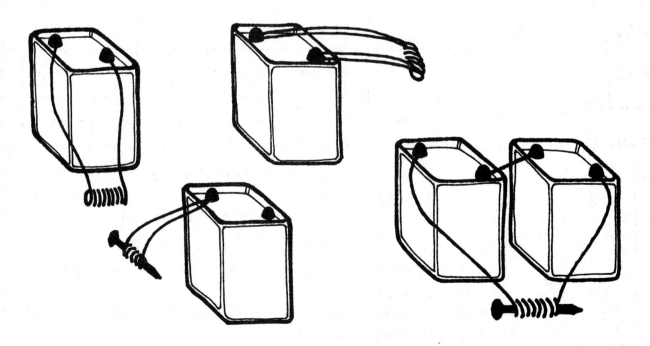

Can you answer Marco Polo's question now? Was the magnet magic? Why did the metal fish always point North? What are a magnet's powers?

THE MAGIC OF MAGNETS

_____

Has successfully mastered THE MAGIC OF MAGNETS

Date: _____

MINI-BEASTS AND CREEPY CREATURES

BACKGROUND

Most mini-beasts used in these experiments are insects. They have an external skeleton made of a tough leathery substance called *chitin*. They have segmented bodies and grow bigger by molting.

In the molting process, the exoskeleton splits open and the creature crawls out. Over the next few hours the new, bigger covering hardens. The creature must constantly bend and unbend its legs during this time to keep them from getting stiff.

Insects breathe mainly through tiny holes called *spiracles*, which are along the sides of their body. The air then moves along tubes as it diffuses through the mini-beast's body.

Arthropods lay hundreds, even thousands, of eggs because so many of the eggs and young are eaten by birds, insects, and other animals. The eggs are generally laid in the fall in special egg sacs that harden, forming a protective shield against the harsh winter weather. The young generally hatch in the spring.

The young go either through complete or incomplete *metamorphosis* as they develop. The main difference is that in incomplete metamorphosis the young hatch looking like adults, except that they are much smaller and lack wings. Most adult stages gain wings as well as functioning sexual organs during the final molt. Grasshoppers and praying mantises are examples of insects that have incomplete metamorphosis.

Butterflies and moths are insects with complete metamorphosis. They develop through four main stages: egg, larva, pupa, and adult. Each stage looks completely different from the other stages.

In butterflies and moths the larvae eat twenty-four hours a day. Many types are very picky eaters and will only eat one type of food. Monarch butterfly caterpillars will eat only milkweed leaves, for example.

The pupal stage is often called the *cocoon.* This is a quiet, resting stage of development.

Adult butterflies and moths are different, as the chart shows.

Butterflies	Moths
1. Club-end antennae	1. Feathery antennae
2. Drink nectar with straw-like proboscis	2. Usually do not eat (short-lived)
3. Hold wings together at rest	3. Wings spread out at rest

Spiders are another type of creepy creature. This chart shows the main differences between insects and spiders.

Insects	Spiders
1. 6 legs	1. 8 legs
2. 3 main body parts	2. 2 main body parts
3. Have compound eyes and simple eyes	3. Have only simple eyes (usually 8)
4. Most do not produce silk	4. All produce silk

Mini-beasts are generally hardy creatures and adjust readily to life in terrarium habitats. Being very set in their routines, they respond with predictable behavior to basic experiments.

Insects and spiders generally prefer warm, moist, dark places to cold, dry, bright habitats. This behavior is also true of earthworms — another creepy creature that is easy to find and interesting to observe.

WORD BOX:

metamorphosis	proboscis	larva
pupa	pupate	environment
cold-blooded	cocoon	

LEARNING OBJECTIVES:

1. Students will be able to identify a number of different mini-creatures.

2. Students will understand the mini-beasts' general requirements for life.

3. Students will understand general insect life cycles and be able to explain complete and incomplete metamorphosis.

4. Students will be aware of how mini-beasts react to the environment.

PREDICTIONS:

1. Do you think mini-beasts will prefer warm or cool places?
2. Do you think mini-beasts will prefer light or dark places?
3. Do you think mini-beasts will prefer dry or damp places?

DISPLAY AND BULLETIN BOARD IDEAS:

1. Make a large grasshopper for the bulletin board. Label its main body parts.

2. Put letters on the bulletin board for COMPLETE METAMORPHOSIS and INCOMPLETE METAMORPHOSIS. Use a butterfly to illustrate complete metamorphosis. Show an egg, a caterpillar, a cocoon, and an adult butterfly. Connect the stages with large arrows to illustrate that this is a cycle. Use a grasshopper to illustrate incomplete metamorphosis. Show three or four grasshoppers all exactly alike (without wings) but each one increasingly bigger. Make one even larger grasshopper with wings. Connect all of these stages with large arrows to illustrate that this is also a cycle.

3. Collect colorful pictures of insects and other mini-beasts and creepy creatures for a bulletin board. Encourage students to bring in pictures.

4. Have a working ant farm and earthworm farm for students to observe. Plastic ant farms are available. Ants and earthworms do well in rich, loose soil in gallon jars.

5. Make a display of books about insects, spiders and other mini-beasts. You may want to include fiction books as well as nonfiction books.

EXTRA SPARK STARTER:
Make mini-creatures together as a class.

BUTTERFLIES

Supplies needed: Construction paper   Scissors
                      Colored tissue paper   Glue
                      Pipe cleaners

How to do it:

1. Draw the body with attached wings on the construction paper. Cut out this shape.

2. Cut sections out of the wing area. Glue tissue paper on the wings to cover these cut-out sections.

3. Add pipe cleaners. Glue them in place.

4. Hang the butterflies from the ceiling so they can fly in the warm air currents in the room.

EGG CARTON CREATURES

Supplies needed: Construction paper   Scissors
                   Glue   Pipe cleaners
                   Egg cartons   Crayons or
                                           felt markers

How to do it:

Let your imagination run wild. Create bright, colorful, fanciful beasties.

MATERIALS NEEDED:

Wire coat hangers
Broom handle
Mosquito netting or
   fine-meshed screen
Plastic lid from
   shortening can
Bird seed
Quart jars
Paper towels
Lamp
Banana
Gauze
Ice
Dark construction paper
Soft paper tissue
Paper or plastic cups

Wire cutters
Electrical tape
Needle and thread
Stapler
Soil
Pop bottle caps
Metal ring lids for
   quart jars
Cotton swabs
Masking tape
Magnifying glass
Spray can of white or
   silver paint
Sugar
Sugar-coated cereal
Measuring cup

Pliers
Scissors
Flower pots
   (medium and big)
Staples
5- or 10-gallon fish
   tank or 1-gallon jars
Epoxy glue
Half-gallon cardboard
   milk cartons
Rocks and sticks
Rubber band
Insect net
Small towel
Shoe box
Modeling clay

Mini-beasts To Be Caught:

Caterpillars   Crickets   Grasshoppers
Butterflies   Any others that
                       interest students

Copyright © 1980 — THE LEARNING WORKS

ADDITIONAL SOURCES: BOOKS

Conklin, Gladys. I Watch Flies. Holiday House, 1977.

Ewbank, Connie. Insect Zoo. Walker & Co., 1973.

Hornblow, Leonora, and Arthur Hornblow. Insects Do The Strangest Things. Random House, 1968.

Oxford Scientific Films. The Butterfly Cycle. G. P. Putnam's Sons, 1977.

Sterling, Dorothy. Insects and The Homes They Build. Doubleday, 1954.

ADDITIONAL SOURCES: FILMSTRIPS

Collecting Insects and Other Small Animals. Intermediate level. Set of 5 filmstrips and sound. Encyclopedia Britannica Corporation.

Helpful Insects. Primary and intermediate level. Set of 5 filmstrips with captions. Encyclopedia Britannica Corporation.

Investigating Insects. Grades 4-6. Set of 6 filmstrips and sound. Coronet.

Life Cycles. Grades 5-12. Set of 5 filmstrips and sound. National Geographic Society.

What Is An Insect? Learning Shelf Kit (pictures, mini-loop, printed text, cassette, duplicating masters, teacher's guide). National Geographic Society.

ANSWERS:

p. 85: 1. The larvae look like small white worms.
2. The food is the banana.
3. The larvae are the most active when the temperature is between 20° and 24°C. They are never completely quiet.

p. 86: 5. The pupal stage darkens, hardens and becomes brown.
6. The pupae are not active.
7. The adults appear about two to three days after the beginning of the pupal stage. The entire cycle from egg to adult requires only about eight or nine days.
8. Answers will vary.

p. 88: The results of the tests will vary. Caterpillars should be particularly sensitive on the hairs along the sides and back.

p. 89: The results will vary. There should be more in the warm jar than in the cold jar. Think about some of the variables that could affect your results.

p. 90: The results will vary, but most mini-beasts like to hide in dark places. Crickets will seek the dark more quickly than grasshoppers.

p. 92: The butterfly should respond best to having the sweet juice touch its head and feet. Like many mini-beasts, butterflies taste through their feet.

EVALUATION:

Number the stages of development of the fruit fly in order.

___(4)___ Adult   ___(1)___ Egg   ___(2)___ Larva   ___(3)___ Pupa

Each statement is incorrect in some way. Cross out what is not true and print above it the word or words that makes the statement correct.

1. Mini-beasts like ~~cold~~ (warm) climates better than ~~warm~~ (cold) ones.
2. Mini-beasts like places that are very ~~dry~~ (damp).
3. Mini-beasts like ~~bright, sunny~~ (dark) places.
4. The ~~larva~~ (pupa) stage is the quiet stage before adulthood.
5. Butterflies can taste through their proboscis and their ~~abdomen~~ (feet).
6. One way to tell a butterfly from a moth is that a ~~butterfly~~ (moth) has feathery antennae.

EXTENDED LEARNING:

1. Have a visit from a bee keeper.

2. Write a story about a day in the life of a mini-creature.

3. Tell about the world as if you were a mini-creature.

4. Write a report about a mini-creature. Use at least two different sources.

5. Start a mini-beast zoo.

6. Build an ant farm.

7. Write about Ms. Muffet's adventures as a spider.

8. Everybody is a mini-creature for one day. Younger students may want to make costumes. Each person gives clues. Can anyone guess what creature they are?

9. For older students, put grasshoppers in alcohol or formaldehyde (can be purchased at a drugstore). Use them to look at the internal structure.

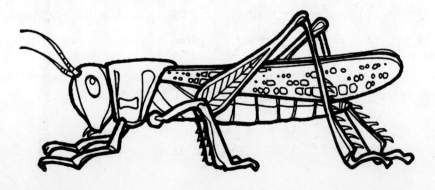

LITTLE MS. MUFFET

Little Ms. Muffet was a cute, young thing. She had an Afro, big brown eyes, and a dimple in her left cheek. She would have been beautiful if it had not been for one thing. She was spoiled.

Muffet had everything that she wanted. She had a complete set of Star Wars characters, every computer game on the market, and ten KISS T-shirts.

Muffet had never had to lift a finger to get what her little heart desired.

Then one night while Muffet was sitting on a red satin pillow eating a hot fudge sundae, a big black spider dropped down beside her.

"Eeeeek!" shrieked Muffet. "Maid, bring the Raid!"

"Hold it, sweetie," said the spider. "I'm not an ordinary spider. I'm your buggy godmother.

"I'm going to show you what the world is really like. I'm going to let you see the world from another point of view. Nobody works harder than the mini-beasts of this world."

With that, the spider shot some silk at Muffet.

MUFFET SHRANK

and

shrank

until she was a little, brown-eyed spider.

What will happen to Muffet? What will she find out about the world from a mini-creature's point of view?

As you do the following experiments, keep Ms. Muffet in mind.

I, TD   Name _____

HOW CAN YOU MAKE AN INSECT NET?

Supplies needed: 1 wire coat hanger   1 yard of 36-inch width
                          Wire cutters   cheesecloth, mosquito
                          Pliers   netting or nylon net
                          1 broomstick handle   Scissors
                          Electrical tape   Needle and thread

How to do it:

1. Cut the hanger on either side of the hook. You may need help.

2. Bend the wire into a hoop shape. Use the pliers to help you.

3. Put the ends of bent wire on either side of the broom handle.

4. Tape the wire in place by wrapping tape over the wire and around the broom handle.

5. Cut out a shape similar to the one shown from the netting.

6. Sew the open edge together.

7. Gather the small opening shut and sew it tightly.

8. Loop the edge of the large opening over the bent wire hoop.

9. Sew this edge with small, even stiches so that the net will stay firmly in place.

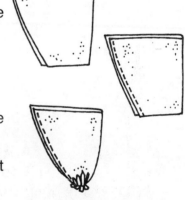

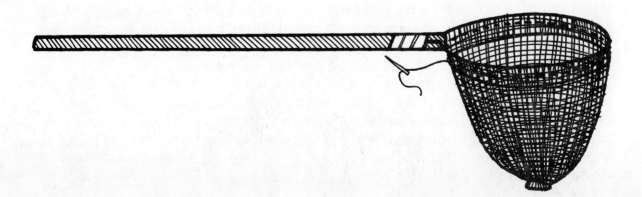

Name _____

HOW CAN YOU MAKE HOMES FOR MINI-BEASTS?

A Flowerpot House

Supplies needed:
1 medium or large flower pot
Soil with weeds or grass
A pop bottle cap
A piece of fine-meshed screen (wide enough to just fit inside the rim of the pot when rolled up)
Stapler
Electrical tape
Plastic lid from shortening can

How to do it:
1. Fill the pot with soil, weeds, and grass.
2. Press the pop bottle cap into the soil and fill it with water. Mini-beasts need to drink water, too.
3. Roll the screen and staple the edges together. Press tape along the seam (both on the inside and on the outside) to make sure that the seam is closed.
4. Put the screen into the pot and put the plastic lid over the top.
5. When you catch insects or other creepy creatures, open the lid and drop them inside.

A Terrarium

Supplies needed:
1 5- or 10-gallon fish tank or a 1-gallon jar
Soil with grass and other plants
Rocks and sticks
Pop bottle cap
Masking tape
Bird seed
Piece of fine-meshed screen or nylon net slightly bigger than the top

How to do it:
1. If this is to be a home for ants, earthworms, or other burrowing creatures, fill about half-way with soil. For spiders, crickets, and other mini-beasts, use less soil.
2. Arrange rocks and sticks for hiding places.
3. Press the pop bottle cap into the soil and fill with water.
4. Make loops of masking tape, press into bird seed and stick to the walls of the terrarium. This adds food for crickets and other seed eaters.
5. For earthworms, include some decaying leaves or peat moss. For spiders, provide some other insects to be food, and if they are web builders, include sticks large enough to support a web.
6. Cover the terrarium with the screen.

EXPERIMENT FOR YOUNG WIZARDS

What is the life cycle of a fruit fly?

As fruit flies develop from eggs to adult flies, they go through a series of changes. These stages of growth are called a *life cycle*. Fruit flies develop through *complete metamorphosis*. The word *metamorphosis* means change. In complete metamorphosis, each stage is completely different. The stages of development are: egg, larva, pupa, and adult.

In this experiment you will be able to see the different stages of the fruit fly's life.

Supplies needed:
- 1 ripe banana
- 1 quart jar
- Magnifying glass or hand lens
- A piece of gauze big enough to fit over the mouth of the jar
- 1 rubber band

How to do it:

1. Put the banana in the clean, dry quart jar. Do not cover it.
2. Let the jar and banana sit near a window or even outside.
3. Check daily. Soon you should find small fruit flies flying around the banana.
4. Use the magnifying glass. Can you see both male and female fruit flies? (Fruit flies are called Drosophila by scientists.)

 Male
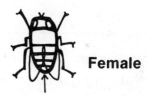 Female

The end of a male fruit fly's body is solid black. The end of a female fruit fly's body is striped.

5. Use the magnifying glass. Do you see any small white eggs on the banana?
6. If you do, cover the jar with the gauze and rubber band. If not, wait another day and check again.

What happens:

1. In a day or two the eggs will hatch. What do the young flies look like? _____

   Young flies are called *larva* (larvae is the plural form of this word).

2. What is the food that these young flies eat? _____
3. Watch the larvae in the morning, afternoon, and night. Check the temperature. When are the larvae the most active? _____

A   Name _____

THE FRUIT FLY'S LIFE CYCLE — CONTINUED

4. After about seven to ten days, the larvae will go into the next stage of development. The larvae will *pupate* — go into a quiet stage as they change to their adult form.

5. How does the appearance of the young flies change during the pupal stage? _____

6. Are the pupae active? _____

7. How soon after the larvae went into the pupal stage do the adult flies appear? _____

8. How many adult males do you see? _____

   How many adult females do you see? _____

Draw a picture of each stage of the female fruit fly's life cycle.

EGG	LARVA	PUPA	ADULT

FUN SPOT

There are many kinds of flies. True flies have only two wings. Other flying insects have four wings. Unscramble the names of these flies and then match the name to the picture.

1. YEHROSLF _____

2. WBOLLYF _____

3. EDMGI _____

4. LAYLGLF _____

BLOWFLY

HORSEFLY
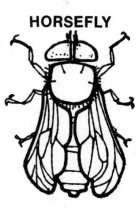

GALLFLY

MIDGE

Name _____

FUN PAGE

You can become a spider web expert and even start a web collection.

To collect webs you need: Large sheets of dark-colored construction paper
A spray can of white or silver paint
Scissors
A friend (if possible)

How to do it:

1. When you see a web that you would like to collect, look around carefully. A surprised web-owner may be hiding nearby.

   Although most spiders would rather run away from people, certain female spiders will try to protect their egg case. Some spiders are also more aggressive than others.

2. Lightly spray the web with spray paint.

3. Press the dark paper evenly against the back of the web.

4. While you hold the paper, have your friend cut the spider's support or bridge lines with the scissors. This frees the web from the branches or leaves around it.

5. Let the web dry on the paper. The outline of the carefully constructed web will be clearly visible.

Spiders instinctively know how to build their webs. They create beautiful geometric patterns by measuring and spacing the silk strands with their back legs.

You can sew your own web. Use light-colored thread on dark paper. The pictures show the step-by-step pattern that Argiope, the orb-web weaver, follows in creating its delicate web. With this web, the spider catches its prey.

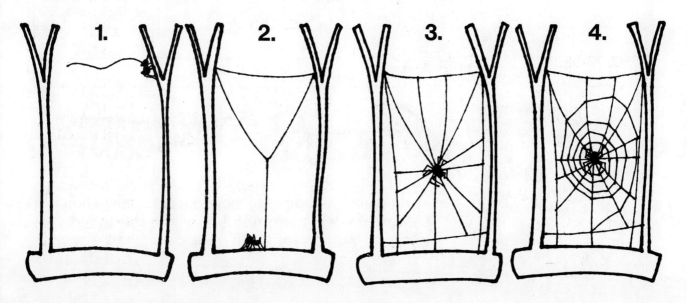

Name _____

IS A CATERPILLAR TICKLISH?

Supplies needed: At least 2 different kinds of caterpillars (more if possible)
Paper towels
Cotton swabs
Mini-beast house with plenty of fresh leaves of the kind where the caterpillar was found

Caterpillars eat twenty-four hours a day. If you are housing your caterpillars, provide plenty of food. Some caterpillars are picky eaters. Gather leaves from the type of plants that were already serving as dinner when you captured your caterpillars.

How to do it:

1. If possible, collect some caterpillars that have long, fuzzy hair as well as smooth-skinned varieties.

2. Handle the caterpillars gently. Scoop them up on a paper towel and avoid touching them, if possible. Caterpillars breathe through their skin. Holding them blocks their breathing area; squeezing may also damage them.

3. Test the caterpillars one at a time.

4. Test each caterpillar by touching it with a cotton swab in the ways shown on the chart and write down everything that the caterpillar does. If a reaction shows up, let the caterpillar return to normal before the next test.

What does the caterpillar look like?	Touch outer hairs or skin along the back	Touch head	Touch tail	Touch side hairs or skin

Why not keep your caterpillars in a mini-beast house with plenty of food until they spin cocoons? Then in a few weeks or in the spring, you will be able to see what kind of butterfly or moth your caterpillars become.

Copyright © 1980 — THE LEARNING WORKS

I   Name _____

DO INSECTS PREFER WARM OR COLD CLIMATES?

Supplies needed:
- Soil and plants
- Measuring cup
- 2 quart jars
- 2 screw-on metal ring lids that fit the jars
- Epoxy glue
- 1 insect net
- 2 half-gallon cardboard milk cartons
- Scissors
- Ice

How to do it:

1. Dig up sod with weeds and grass. Scoop up about four cups of soil.
2. Place the quart jars on their sides. Fill each jar about half full of soil and plants.
3. Attach the two lids together with the glue.

4. Use the insect net to catch a number of small crickets or grasshoppers.
5. When the lids are dry, put the insects inside both jars and screw the two jars together.
6. Use the scissors to cut the milk cartons so that a cradle is created for each jar.

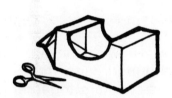 

7. Pack ice around one of the jars.

What happens:

Observe carefully and complete the chart. Count how many insects you see in each jar at each time interval.

TIME	WARM JAR	COLD JAR
15 minutes		
30 minutes		
1 hour		
2 hours		

Puzzling out the results:

Which kind of climate do the insects like best?

Insects — like most mini-beasts — are cold-blooded. This means that their body temperature is nearly the same as their environment.

DO INSECTS PREFER LIGHT OR DARK PLACES?

Use what you learn in this experiment. Observe other creepy creatures during the day and at night. Watch for them with a flashlight and around a porch light. Cut out each mini-beast picture at the bottom of the page and paste it in the column that shows when you saw that creature.

DAY	NIGHT

Supplies needed:
- Soil and plants
- 2 quart jars
- 2 screw-on metal ring lids to fit the jars
- Epoxy glue
- 2 half-gallon cardboard milk cartons
- Scissors
- 1 insect net
- A small towel
- A lamp

How to do it:

1. Set up the jars, lids and milk carton cradles by following the directions on page 89.
2. Catch a number of small crickets or grasshoppers with the net. Put an equal number of mini-beasts in each jar. Screw the two jars together.
3. Cover one jar with the towel so that the inside is dark.
4. Set up the lamp so that the other jar is continuously lighted. Be sure that the light is not too close to the jar. The light will make the jar hot.

What happens:

Observe carefully and complete the chart. Count how many insects you see in each jar at each time interval.

TIME	LIGHT JAR	DARK JAR
30 minutes		
2 hours		
Next day		

Earthworm

Mosquito

Butterfly

Grasshopper

FUN PAGE

Can you find a super hopper?

Catch several grasshoppers.

One at a time, have a grasshopper leap from a starting line.

Mark where the grasshopper lands in one jump.

Which grasshopper is the super hopper?

Have races with your friends.

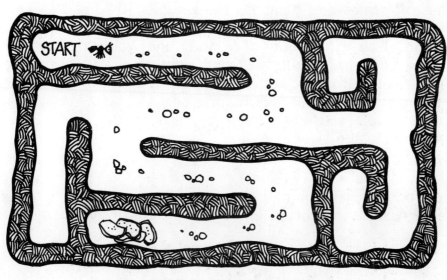

How smart are ants?

Supplies needed: A shoe box   Modeling clay
                   Sugar-coated cereal

Build a simple course inside the shoe box, making walls out of clay. Crumble several flakes of sugar-coated cereal and sprinkle a path to the grand reward — a pile of cereal. Put an ant at START. Can the ant follow the path to the reward? Time the ant. How many seconds — minutes — does it take the ant to find its reward? Is one ant faster than another? Climbing the walls is cheating and disqualifies the ant.

WHERE DO BUTTERFLIES TASTE?

Supplies needed: Insect net
Soft paper tissue
Magnifying glass
Paper or plastic cup
3 tablespoons of water
2 tablespoons of sugar
1 cotton swab
Mini-beast house (if butterflies are to be kept)

How to do it:

1. Catch several butterflies. Do not catch moths. Many moths do not eat after the caterpillar stage. How can you tell a moth from a butterfly? Look at the pictures of a resting moth and butterfly. Watch for the wing position and the antennae shape.

Moth   **Butterfly**

2. Mix the sugar and water in the cup to make a sweet syrup.
3. Gently hold the butterfly to be tested with its wings together.
   Use soft paper tissue to hold the butterfly. A butterfly's wings are covered with tiny scales. Look at a wing with the magnifying glass. Rubbing the wing with your fingers brushes off scales and damages the wing.
4. Dip the cotton swab into the sugar water.
5. Touch the cotton swab to the butterfly's body and watch the butterfly's head. A butterfly has a straw-like mouth called a *proboscis*. With its proboscis, the butterfly drinks nectar from flowers. Whenever the butterfly tastes anything sweet — nectar is sweet — it unrolls its proboscis.
6. Gently touch the cotton swab to each part of the butterfly's body. Between each test, redip the swab in the sweet water.

What happens:

Use crayons, colored pencils or markers to circle on the picture below any parts where the butterfly has a sense of taste. Make the butterfly very colorful.

Muffet's buggy godmother turned her back into a little girl.

Muffet was never quite the same again. Among other things, little Ms. Muffet always found herself getting hungry when she saw a fly.

What did Muffet learn about mini-beasts and creepy creatures?

MINI-BEASTS AND CREEPY CREATURES

_____

**Has successfully mastered
MINI-BEASTS AND CREEPY CREATURES**

Date: _____

AIR: THE INVISIBLE FORCE

BACKGROUND

Air is an invisible substance that we live with and seldom take any notice of. Yet air is essential to life. Air presses down on every square inch of our bodies. The force of air carries us into the sky to join the winged creatures. The force of air can be strong enough to rip trees from the ground or blow apart buildings.

Galileo Galilei was one of the first to investigate the properties of air. He proved that air takes up space, has weight, and has force.

To understand air completely, it is necessary to realize that air is really made up of moving atoms and molecules of gas. These particles are widely spaced even at the earth's suface, but the higher up you go the more widely spaced they become.

The blanket of gas around our earth is called our *atmosphere*. At the surface of the earth, the air presses down on us at more than fourteen pounds per square inch. On high mountains this pressure is much less — only five and a half pounds per square inch.

Most of the earth's air is very near the earth's surface, and yet even the thin outer layer is important to us. Our atmosphere blocks out most of the cosmic rays from outer space. The radiation would be deadly if it all hit the earth.

Friction against air particles causes meteors — chunks of metal and stone from space — to burn up before they reach the ground. Only the remains of the biggest meteors ever crash into the earth. Without the air, our earth would be covered with craters from the impact of meteors, just like the moon.

Air molecules and atoms also help to heat the earth and circulate the air. The sun's rays, striking air particles and dust, give off heat. As the air is warmed, the particles begin to move faster and spread farther apart. The warm, lighter air rises. The cool, heavier air sinks. The mixing of these two creates wind.

People have learned to use the force of air to do work. Air pressure has been used to lift heavy weights.

The Wright brothers experimented first with kites, and their early flying machines were a modified type of Hargrave box kite. They learned what shape could make the most of the air's lifting power with the least weight. People have also learned that channeled air can create a strong thrust, as in the rocket balloon experiment (p. 106).

Daniel Bernoulli discovered that in an airstream faster moving air has lower pressure than slower moving air. Today's planes have wings that are curved on top and flat underneath. In flight, the air that moves over the wing goes farther, and faster, than the air that goes under the wing. The higher pressure of the slower air under the wing pushes up on the wing as it moves toward the lower pressure above. This gives the plane lift.

WORD BOX:

inflate   deflate   air pressure

LEARNING OBJECTIVES:

1. Students will be able to state the main properties of air: air has weight, air takes up space, and air has force.

2. Students will understand that air rises when it is heated.

3. Students will be able to use the properties of air to do work.

4. Students will understand that fast-moving air has less weight than more slowly moving air.

PREDICTIONS:

1. You can't see air. Do you think air takes up space?

2. Which do you think weighs more, a ball full of air or an empty (deflated) ball?

3. Can you hold a glass full of water upside down without the water falling out?

4. Why do you think that airplane wings are curved on top and flat on the bottom?

5. Why do you think airplanes take off into the wind instead of with the wind behind them?

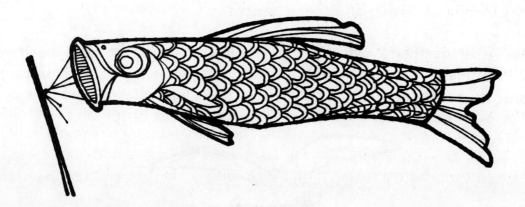

DISPLAY AND BULLETIN BOARD IDEAS:

1. Write to NASA
   Code LFG-13
   Washington, DC 20546

   Ask for free booklets and pictures about the history of flight and space exploration. They also have free posters about the space program that make exciting and colorful room decorations. Write early to allow time to receive your materials.

2. Create a bulletin board display of pictures that show air in action: wind blowing, moving sail boats, kites riding air currents, gliding birds, flags blowing.

3. Set aside an area on the bulletin board for challenge questions. These may include such ideas as:

   1. Why do you suppose a flag won't flutter in the breeze on the moon?

   2. Why do you sunburn more easily on top of a mountain than at sea level?

   3. Why do you think tornadoes happen mainly during hot weather rather than during cold weather?

   Encourage your students to write their responses. Display the most interesting, creative, and accurate ideas.

4. Have a mobile made of colorful papier-mâché or posterboard balloons. Each balloon could also state a fact about air.

EXTRA SPARK STARTER:

Give everyone a balloon. Have them blow it up. Tell them that they are now holding what they are about to study. What is it?

Using the balloon, make a chart together of all the qualities of air that you and your students can think of. Listen to the air in the balloon, squeeze the balloon gently. Examine air, using all of your senses.

Make a list together of all the questions your students can think of that they would like to investigate. Partway through the unit, reexamine this list. Discuss the things that are on the list and have already been investigated. Add any new questions that may be raised. Consider the list again at the end of the unit.

MATERIALS NEEDED:

Glasses	5- or 10-gallon fish tank	Measuring cup
Balloons	Ice	Hot pad
Empty 3-pound can	Modeling clay	Gallon milk jug or large
Pot holder	Quart jar	metal can and lid
Long plastic straws	Scissors	Monofilament fishing line
Food coloring	Masking tape	Large garbage bag
Meter stick	Hole punch	Wooden dowels about
Chalk	Mirror	6 mm thick
Notebook paper	Ping-Pong ball	Kitchen scales or triple
Blow dryer	Paper clip	beam balance scale
Steel tape measure	Electric skillet	Index card
	Empty pop bottle	Inflatable ball

ADDITIONAL SOURCES: BOOKS

Asimov, Isaac. Great Ideas of Science. Houghton Mifflin, 1969.

Goldstein-Jackson, Kevin. Experiments With Everyday Objects. Prentice-Hall, Inc., 1978.

Stone, Harris, and Bertram M. Siegel. Puttering With Paper. Prentice-Hall.

Stone, Harris, and Bertram M. Siegel. Take A Balloon. Prentice-Hall.

White, Lawrence B., Jr. Investigating Science With Paper. Addison-Wesley Publishing Co., 1970.

Wyler, Rose, and Eva-Lee Baird. Science Teasers. Harper & Row, 1966.

ADDITIONAL SOURCES: FILMSTRIPS

Investigating Heat. Grades 4-8. Set of 4 filmstrips and sound. Coronet.

ANSWERS:

p. 101: What happens:
    1. No.
    2. Yes — the air keeps the water out.

Copyright © 1980 — THE LEARNING WORKS

p. 101: 3. When tipped, the air escapes as bubbles and the water replaces it inside the glass.
Puzzling:
Air bubbles will push out the water.

Air does take up space. We know this because the water and the air could not fill the glass at the same time. The air pushed out the water.

p. 102: What happens:
The balloon inflates when the air inside the bottle is heated. It deflates as the air cools.

Puzzling:
When the air is heated, it rises and the molecules move farther apart. When the air is cooled, it sinks and the molecules move closer together.

p. 103: Answers will vary, but the inflated ball will weigh more — air has weight.

p. 104: The jug will slowly cave in.
Puzzling:
After capping, there was more air outside the jug than inside. The air inside escaped as it was heated. The jug collapsed due to the greater external air pressure.

p. 105: The water will rise in the straw because there is more air pressure pushing down on the colored water in the jar than there is inside the bottle.

A bigger and more powerful fountain could be made by:
1. Using a bigger container of colored water
2. Heating the bottle longer to drive out most of the air
3. Using a narrower tube to increase the force of the rising water.

Also a tighter seal around the straw would help.

p. 106: The results will vary. The rocket can be improved by using a more oval balloon or a balloon with a narrower opening, or changing the place where the straw is attached.

p. 110: The plane will fly better if thrown into the wind.
More weight on the nose is also helpful.

p. 111: The fast-moving air of your breath has less weight than the air below the paper. The paper strip will rise as you blow down on it.

The Ping-Pong ball will remain suspended in the air shooting up from the blow dryer. The fast-moving air has less weight under the ball than the air above the ball.

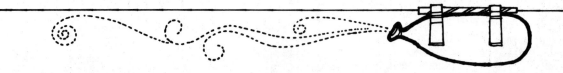

EVALUATION:

Choose the letter of the word or words that correctly completes each statement. Print the letter on the blank.

A. Less
B. Rises
C. Inflated
D. Air pressure
E. Hot and cold air
F. More
G. Faster
H. Slowly
I. Deflated
J. Water and salt

1. When air is heated it __(B)__.

2. Wind is caused by the mixing of __(E)__.

3. An inflated ball weighs __(F)__ than a deflated ball.

4. Bernoulli discovered that fast-moving air weighs less than more __(H)__ moving air.

5. Air moves __(G)__ over a curved surface than over a flat surface.

6. When a balloon is full of air, we say that is __(C)__.

7. __(D)__ is the weight of the air pushing against something.

Write the three main properties of air.

1. (Air has weight.)
2. (Air takes up space.)
3. (Air has force.)

EXTENDED LEARNING:

1. Organize an airplane contest. Keep the materials all the same. Challenge your students to design the best paper airplane that they can using those materials.

   Graph the flight results either individually or as a class.

2. String a clothesline across the room. Make pictures as a class that illustrate important moments in the history of flight. Clip the pictures to the line.

3. Find out more about the Wright brothers, Bernoulli, Lindbergh, Rogallo, Goddard and others involved in the history of flight.

4. Have a pilot visit.

5. Get flight schedules and flight plans from a major airline. Draw the routes on a map. Use a map scale to figure how many miles the plane will fly from city to city and in a complete flight. Use arrival and departure times to figure the time spent in the air.

Copyright © 1980 — THE LEARNING WORKS

THE TEST FLIGHT

Wilbur Wright straightened up from the glider resting on the crest of the sand dune. The satin cloth pulled tight across the glider's frame glistened in the sunlight. "I hope this front elevating rudder will give the glider better balance," he said.

"Let's find out if it works," his brother, Orville, answered.

Wilbur stretched out on the lower wing while Orville and their friend William Tate each took one of the ropes attached to the wings. Pulling on the ropes, William and Orville started down the dune — trotting at first and then running until the gusty wind lifted the glider.

"Let go," Wilbur yelled.

When Wilbur pulled the cords to raise the rudder, the glider climbed to nearly thirty feet.

Suddenly, one of the rudder ropes snapped in his hand. The glider nosed down and like a broken kite smashed into a sand dune.

Orville and William ran to the wreckage. Wilbur wasn't hurt. He sat up still holding onto the one rudder rope.

"It needs more work," Wilbur announced.

In the following experiments, you will investigate the invisible force — air — just as the Wright brothers did.

Do you think the Wright Brothers can really build a flying machine?

I Name _____

DOES AIR TAKE UP SPACE?

Supplies needed: 2 glasses   A 5- or 10-gallon tank

How to do it:

1. Fill the tank about two-thirds full of water.

2. Push one glass straight down into the water, holding it upside down.

What happens:

1. Does the water move up into the glass? _____

2. Can you push the glass to the bottom of the tank without any water getting inside the glass? _____

3. What happens if you tip the glass slightly while it is under water? _____
_____

Puzzling out the results:

Can you catch the air from the glass?

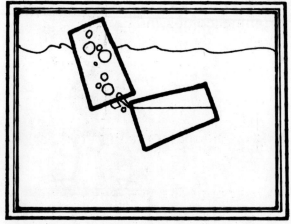

1. Push one glass under the water and let it tip to fill with water.

2. Push a second glass straight down into the water. This glass should be deeper than the one full of water.

3. Hold the two glasses close together.

4. Tip the glass full of air so that the escaping bubbles go into the glass full of water.

What happens when the air bubbles go into the water-filled glass?
_____

Does air take up space? _____

How do you know? _____

I   Name _____

WHAT HAPPENS WHEN AIR IS HEATED?

Supplies needed: 1 empty glass pop bottle   Measuring cup
A balloon that fits over   An electric skillet
top of the bottle

How to do it:

1. Put the balloon over the top of the bottle.

2. Put two or three cups of water in the electric skillet and set the temperature at 250° F. or medium.

3. When the water begins to boil, put the bottle in the middle of the pan.

4. Keep adding water if necessary so that the pan does not become dry. Be sure to unplug the electric skillet when you are finished.

What happens:

    What happens to the balloon after about five minutes?

WHAT HAPPENS WHEN AIR IS COOLED?

Supplies needed: The bottle with the inflated balloon
1 3-pound can such as shortening comes in
Ice

How to do it:

1. Put the bottle with the inflated balloon from

    the experiment above into the can.

2. Pack ice around the bottle.

What happens:

    What happens to the balloon after about five minutes?

Puzzling out the results:

What do you think happens to air when it is heated?

What do you think happens to air when it is cooled?

The mixing of warm and cool air makes wind. Understanding these air currents was important to Wilbur and Orville Wright.

Copyright © 1980 — THE LEARNING WORKS

DOES AIR HAVE WEIGHT?

Supplies needed: An inflatable ball (such as a beach ball)
Kitchen scales or triple beam balance scale
A glass
An index card

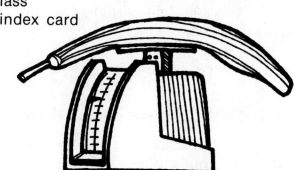

How to do it:

1. Weigh the flat ball on the scales. How much does it weigh? _____

2. Now blow up the ball and weigh it again. How much does it weigh now? _____

3. How much more did the inflated ball weigh than the flat ball? _____

Puzzling out the results:

What made the inflated ball heavier than the flat ball?

_____
_____
_____

FUN SPOT

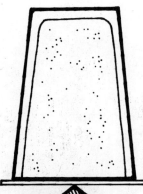

Can you hold a glass of water upside down without the water falling out? You can with the help of the invisible force — air.

Fill a glass full to the very top with water.

Cover the glass with a piece of cardboard or an index card. It must be bigger than the opening of the glass.

Put one hand on the card and one hand on the bottom of the glass. Quickly, turn the glass over and take your hand away from the card.

The water won't pour out.

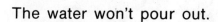

The air pushing up on the card is greater than the weight of water.

I   Name _____

IS AIR REALLY HEAVY?

Supplies needed: Measuring cup
An electric skillet
1 large metal can with a lid (such as ditto fluid comes in) or a plastic gallon milk jug and cap
1 pot holder
1 hot pad

How to do it:

1. Put about two cups of water into the electric skillet. Set the dial for medium or 250° F.

2. When the water is boiling, put the can or jug into the water. Keep adding water if necessary. Do not let the pan go dry.

3. When the can or jug feels warm, put the lid on tight.

4. Lift it from the boiling water with the pot holder and place it on the hot pad. Unplug the electric skillet.

What happens:

What happens to the can or jug as it cools?

Puzzling out the results:

When you capped the can or jug, where was there more air — inside or outside the can or jug? _____

Remember, hot air rises.

What do you think caused what happened to the can or jug? _____

_____

FUN SPOT

Wilbur got lost in the fog. Help him find his way back to camp.
You must pass every sea gull once, but you cannot cross your own path.

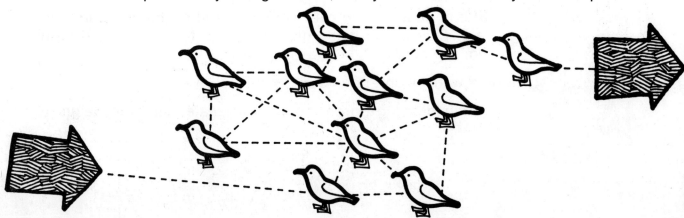

Copyright © 1980 — THE LEARNING WORKS

A

Name _____

EXPERIMENT FOR YOUNG WIZARDS

You can use air pressure to make a fountain in a bottle.

Supplies needed:
Electric skillet
Measuring cup
*1 glass pop bottle
1 quart jar or plastic freezer container
Food coloring
A pot holder
*1 long plastic straw

*Small ball of modeling clay or salt dough (mix equal parts of salt and flour, add just enough water to make dough)

If available, use a flask, a black rubber stopper with a hole that fits the flask snugly, and a long hollow piece of glass tubing in place of the 3 * items.

How to do it:

1. Put about two cups of water in the skillet and set it at 250° F. or medium. Let the water start to boil.
2. Put the bottle or flask in the boiling water. Add more water if necessary to keep the skillet from going dry.
3. Let the bottle remain in the boiling water for about three minutes.
4. Fill the quart jar or plastic container with cool water. Add several drops of food coloring to the water.
5. Stick the straw through the clay. Be sure to squeeze the extra clay out of the straw or snip off the plugged end with scissors. Or, if available, slide the long glass tube through the rubber stopper.
6. Use the pot holder to lift the bottle out of the water.
7. Quickly plug the bottle's opening by sticking the clay and straw (or stopper and glass tube) snugly into the top.
8. Turn the bottle upside down so that the end of the straw (or tube) is in the colored water.

What happens:

WATCH! What happens to the colored water?

Why do you think this happens?

Think of three ways that you could make an even bigger and more powerful fountain.

_____

_____

Copyright © 1980 — THE LEARNING WORKS

| Name _____

FUN PAGE

When rockets blast off into space, they use fuel.
You can use the invisible force — air — to launch a rocket.

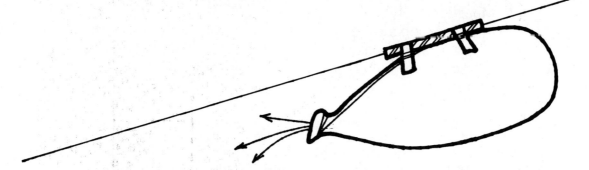

Supplies needed:　　1 straw (plastic is best)　　1 long or oval balloon
　　　　　　　　　　Scissors　　　　　　　　　　Masking tape
　　　　　　　　　　3 meters of monofilament　　A friend
　　　　　　　　　　　fishing line　　　　　　　　Meter stick

How to do it:

1. Cut off an eight-cm piece of the straw and slide it onto the three-meter long line.

2. Tie the fishing line so that it is stretched across an open area of the room (be sure that it is where people will not walk into it). One end should be attached near the ceiling and the other end should be attached near the floor.

3. Blow up the balloon. Hold the end closed with your fingers.

4. Have a friend tape the balloon to the straw. Point the nose of the balloon rocket toward the ceiling.

5. Take the balloon to the low point of the fishing line. Do not let the balloon touch the floor.

　　　　10　9　8　7　6　5　4　3　2　1　　BLAST OFF!
　　　　　Let go of the end of the balloon.

What happens:

How far does your rocket go? (Measure with the meter stick.) _____

Let your friend build a balloon rocket. How far does your friend's rocket go? _____

How could you make your rocket even better? _____

___

To receive free booklets about rockets, write to:
　　　NASA
　　Code LFG-13
　Washington, DC 20546

To write to the astronauts, send your letters to:
　　Astronaut Office, Code CB
　　Johnson Space Center
　　Houston, Texas 77058

Copyright © 1980 — THE LEARNING WORKS

A

Name _____

EXPERIMENT FOR YOUNG WIZARDS

Wilbur and Orville Wright did a lot of experimenting with kites and gliders as they worked to develop their airplane.

Here's a super kite that you can build to experiment with flight on your own.

Supplies needed:
- 1 large garbage bag
- Meter stick
- Scissors
- 1 piece of chalk
- 3 wooden dowels (about 6 mm thick)
- 1 roll of masking tape
- Hole punch
- 1 roll of monofilament fishing line

How to do it:
1. Spread the garbage bag flat. (Double thickness.)
2. Measure and cut the plastic so that the piece remaining measures 90 cm by 50 cm.
3. On the open edge, measure over 25 cm from the upper corner and mark that point. Measure down 25 cm from the upper, open corner and mark. Use the meter stick to help you cut straight between the two marks.
4. Measure in 25 cm from the lower, open corner. Cut straight from that point to the bottom of the upper corner.

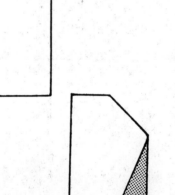

5. The kite will fly better if it has a vent (opening). On the fold edge, measure up 15 cm from the lower edge. Mark this point with the chalk.
6. Measure up 25 cm from the first mark and mark this new point.
7. Cut in 15 cm from the upper mark and straight down to the lower mark.

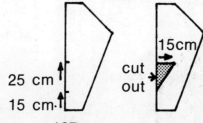

KITE — CONTINUED

8. Open the plastic.

9. Tape the three wooden dowels to the plastic as shown. Use a lot of small pieces of tape. Many small pieces of tape help to strengthen the kite and spread out the pressure of the wind.

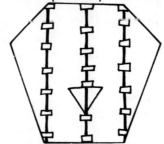

10. Put masking tape on the two upper corners to make them stronger.

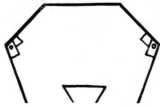

11. Punch a hole in the middle of the tape on each corner.

12. Cut a piece of monofilament fishing line three meters long. Tie one end through each of the corner holes to form the kite's bridle.

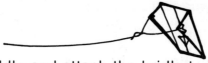

13. Make a loop in the middle and attach the bridle to the rest of the fishing line on its reel.

How high will your kite fly?

Have a kite race with your friends.

To see a picture of the Wright Brother's early glider, connect the dots. Then color the picture.

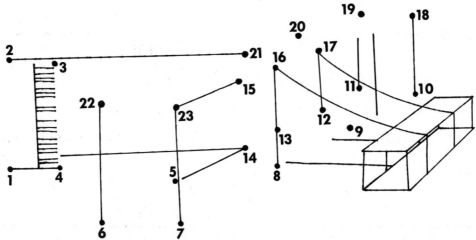

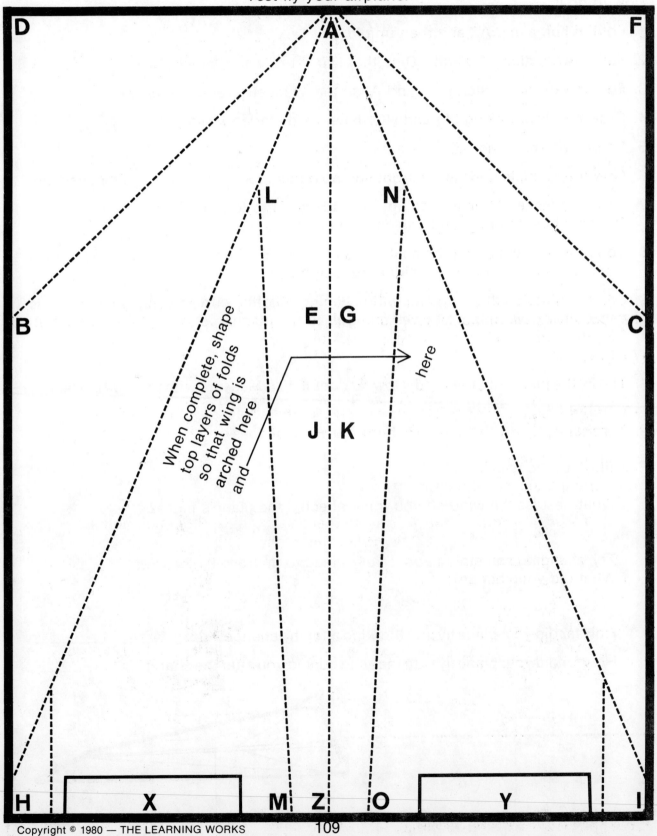

I, TD   Name _____

AIRPLANE — CONTINUED

Supplies needed: Scissors   Steel tape measure
                   Paper clip

How to do it:

1. Fold in half along AZ and then unfold.

2. Fold inward along AB and AC so that D meets E and F meets G.

3. Fold inward again along AH and AI so that B meets J and C meets K.

4. Crease outward along LM and NO (letters are hidden inside).

5. Refold inward along AZ.

6. Grip the multiple folds at the front and hold them together with a small paper clip.

7. You may make optional folds upward at H and I to form rudders. Rudders help to guide and balance the plane's flight.

8. You may also cut along the solid lines at X and Y. Then crease these up or down (try each possibility) for additional rudder control.

9. See the hints on the wings for wing shaping. Slightly curved wings make what is called an *airfoil*. An airfoil gives more lift to the plane.

What happens:

1. Throw the plane into the wind. How far does it go? (Measure with the tape measure.) _____

2. Throw the plane with the wind behind you. How far does it go? _____

Puzzling out the results:

What besides the wind do you think affected the plane's flight?

Try changing something about your plane to improve its flight. What did you change?

How far did your plane fly into the wind after the change? _____

How far did your plane fly with the wind behind you after the change? _____

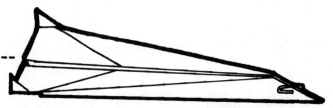

Copyright © 1980 — THE LEARNING WORKS

DOES FAST-MOVING AIR HAVE AS MUCH WEIGHT AS SLOW-MOVING AIR?

Years after Wilbur and Orville Wright developed the airplane, Daniel Bernoulli experimented with air. His goal was to design a better airplane wing. You can repeat some of Daniel Bernoulli's experiments.

PAPER TEST

Supplies needed: A strip of notebook paper the full length of the paper and 2.5 cm wide
A mirror

How to do it:

1. Hold the paper strip by the upper edges.

2. Place the top edge of the paper strip on the outer edge of your lower lip.

3. While looking in the mirror, blow down on the strip. Blow as hard as you can.

What happens to the paper strip when you blow on it?

BIG WIND TEST

Supplies needed: A blow dryer   A Ping-Pong ball

How to do it:

1. Turn on the blow dryer and point it straight up into the air.

2. Hold the Ping-Pong ball in the column of air about ten cm above the top of the blow dryer.

3. Let go of the ball.

What happens to the ball?

Daniel Bernoulli discovered that fast-moving air has less pressure than slower moving air, and that air always follows the path of least resistance.

Daniel Bernoulli also discovered that air moves faster over a curved surface than over a flat surface.

Why do you think he designed an airplane wing that was curved on top and flat on the bottom?

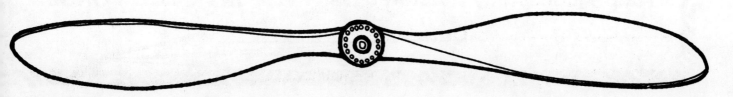

On December 17, 1903, Wilbur and Orville Wright flew the first successful engine-powered machine.

This flying machine (the word *airplane* had not yet been used) rose only three meters and stayed up twelve seconds on its first flight. Short as it was, that was still man's first step toward conquering the air.

What did Wilbur and Orville Wright have to know about air to be able to build the first airplane?

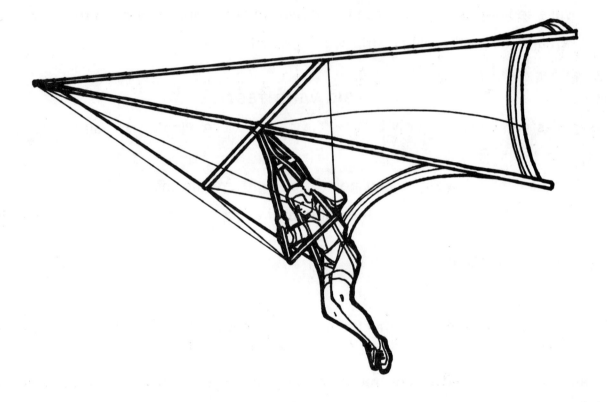

AIR: THE INVISIBLE FORCE

_____

Has successfully mastered AIR: THE INVISIBLE FORCE.

Date: _____